Intermediate Algebra

HARPERCOLLINS COLLEGE OUTLINE

Intermediate Algebra

Joan Dykes, Ph.D.
Edison Community College

HarperPerennial
A Division of HarperCollins*Publishers*

 For information address HarperCollins Publishers, Inc., 10 East 53rd Street, New York, N.Y. 10022.

An American BookWorks Corporation Production

Project Manager: William R. Hamill
Editor: Robert A. Weinstein

Library of Congress Catalog Card Number: 91-55387
ISBN: 0-06-467137-2

92 93 94 95 96 ABW/RRD 10 9 8 7 6 5 4 3 2 1

Contents

Preface

The content and problems contained in *Intermediate Algebra* are similar to the material in a second level developmental college algebra course. Each topic is introduced in a clear, concise manner, followed by examples and solutions.

The solutions are presented one step at a time, with written explanations beside each step. Seeing the steps and reading about the processes will help reinforce the writing and thinking patterns necessary to succeed in algebra. It is assumed that you have access to a textbook which includes the theory behind the steps and shortcuts presented in this guide. Problem sets and answers appear at the end of each chapter to provide practice and immediate feedback.

1

The Real Number System

1.1 BASIC DEFINITIONS

Arithmetic Operations

We will use many symbols and terms throughout our study of algebra. This section reviews some symbols and terms you encountered in beginning algebra.

The four basic operations, addition, subtraction, multiplication, and division are used throughout algebra.

Operation	Example	English Phrases
Addition	$x + y$	The sum of x and y.
Subtraction	$x - y$	The difference of x and y.
Multiplication	xy, $x \cdot y$, x(y), (x)y, (x)(y)	The product of x and y.
Division	x ÷ y, $\frac{x}{y}$, $y\overline{)x}$	The quotient of x and y.

We also use exponents to write repeated multiplications.

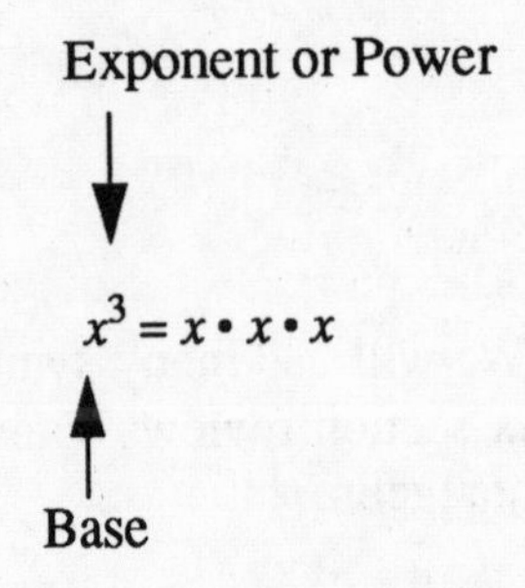

The following symbols are used to group operations together.

Symbols	Name of Symbols
( )	Parentheses
[ ]	Brackets
{ }	Braces
| |	Absolute Value Bars
———	Fraction Bar

We use the following equality and inequality symbols to compare quantities:

Symbol	Example	English Translation
$<$	$2 < 4$	2 is less than 4
$>$	$6 > 1$	6 is greater than 1
$\leq$	$3 \leq 5$	3 is less than or equal to 5
$\geq$	$8 \geq 8$	8 is greater than or equal to 8
$=$	$x = y$	x is equal to y
$\neq$	$x \neq z$	x is not equal to z

EXAMPLE 1 Translate each statement into symbols.

a) The sum of x and 4 is less than 8.

b) The difference of 10 and 8 is greater than or equal to 0.

c) The product of 5 and x is 25.

d) The quotient of d and 8 is less than or equal to 24.

e) The product of 5 and x squared is 4 more than the sum of x and 3.

SOLUTION 1

a) $x + 4 < 8$ — Sum means to add.

b) $10 - 8 \geq 0$ — Difference means to subtract.

c) $5x = 25$ — "is" means is equal to.

d) $\frac{d}{8} \leq 24$ — Quotient means to divide.

e) Do this translation a phrase at a time:

The product of 5 and x squared:

$5 \cdot x^2$ or $5x^2$

4 more than the sum of x and 3

$4 + (x + 3)$

Join the two phrases together with an equality symbol to replace "is"

$5x^2 = 4 + (x + 3)$

Order Of Operations

The grouping symbols are often used to indicate where to start. However, consider an expression like

$$4 + 2 \cdot 8$$

where there are no grouping symbols. You might be inclined to add first, then multiply, giving you an answer of 48. If you try this on your calculator, you will get an answer of 20. Mathematicians have agreed upon the following order of operations.

Order of Operations

1. Simplify inside grouping symbols, starting with the innermost pair and working out.
2. Simplify exponents.
3. Perform multiplication and division steps in order from left to right.
4. Perform addition and subtraction steps in order from left to right.

Thus our example $4 + 2 \cdot 8$ should be done as

$4 + 2 \cdot 8 = 4 + 16$ Multiply: $2 \cdot 8 = 16$

$= 20$ Add.

You might find it helpful, in remembering the order of operations, to memorize the expression: PEMDAS, or Please Excuse My Dear Aunt Sally, which stands for Parentheses, Exponents, Multiplication, Division, Addition, Subtraction.

EXAMPLE 2 Simplify each expression.

a) $4 + 3(2 + 5)$

b) $3 \cdot 5^2 - 4 \cdot 3^2$

c) $5 + 16 \div 2 \cdot 3$

d) $3 + 4[12 - 2(4 - 1)]$

SOLUTION 2

a) $4 + 3(2 + 5) = 4 + 3(7)$ — Work inside parentheses.
$= 4 + 21$ — Multiply: $3(7) = 21$.
$= 25$ — Add.

b) $3 \cdot 5^2 - 4 \cdot 3^2 = 3 \cdot 25 - 4 \cdot 9$ — Simplify exponents: $5^2 = 25$, $3^2 = 9$
$= 75 - 36$ — Multiply before subtracting.
$= 39$

c) $5 + 16 \div 2 \cdot 3 = 5 + 8 \cdot 3$ — Multiply and divide in order from left to right: $16 \div 2 = 8$.
$= 5 + 24$ — Multiplication is next: $8 \cdot 3 = 24$.
$= 29$ — Add.

d) $3 + 4[12 - 2(4 - 1)] = 3 + 4[12 - 2(3)]$ — Work inside parentheses.
$= 3 + 4[12 - 6]$ — Work inside brackets. Multiply: $2(3) = 6$.
$= 3 + 4[6]$
$= 3 + 24$ — Multiply: $4[6] = 24$.
$= 27$ — Add.

Sets

Sets can be used to categorize types of numbers and types of solutions. A set is a collection of objects or things. The objects within the set are called **elements**. Sets are written in braces, and are usually named with capital letters. So,

$$A = \{0, 1, 2, 3, 4\}$$

is the set A containing elements 0, 1, 2, 3, 4. You need to be familiar with the following set notation:

Symbol	Meaning	Example
$\in$	is an element of	$4 \in \{0, 1, 2, 3, 4\}$
$\notin$	is not an element of	$5 \notin \{0, 1, 2, 3, 4\}$
$\subseteq$	subset	$\{0, 1\} \subseteq \{0, 1, 2, 3, 4$
$\nsubseteq$	is not a subset	$\{5, 6\} \nsubseteq \{0, 1, 2, 3$
$\cup$	union	$\{1, 2, 3\} \cup \{3, 4, 5\}$ $\{1, 2, 3, 4, 5\}$
$\cap$	intersection	$\{1, 2, 3\} \cap \{3, 4, 5\}$ $\{3\}$

In the examples above we have listed the elements in the sets. When a list is too long or too tedious, we use **set-builder notation**:

$$\{x \mid x \text{ is an even integer}\}$$

which is read "the set of all x such that x is an even integer." (The bar stands for "such that".) Another technique used to write sets uses three dots (. . .) to indicate that the set continues in the same pattern:

$$\{2, 4, 6, 8, \ldots\}$$

EXAMPLE 3

Let $A = \{0, 1, 2, 3\}$, $B = \{0, 2, 4\}$ and $C = \{1, 3, 5\}$.

Find each set.

a) $A \cup B$

b) $A \cap B$

c) $\{x \mid x \in A \text{ and } x > 1\}$

d) $\{x \mid x \in A \text{ and } x \notin C\}$

e) $B \cap C$

SOLUTION 3

a) $A \cup B = \{0, 1, 2, 3\} \cup \{0, 2, 4\}$

$= \{0, 1, 2, 3, 4\}$ — List elements in *A or B*.

b) $A \cap B = \{0, 1, 2, 3\} \cap \{0, 2, 4\}$

$= \{0\}$ — List elements in *A and B*.

c) $\{x | x \in A \text{ and } x > 1\} = \{2, 3\}$ List elements in A that are greater than 1.

d) $\{x | x \in A \text{ and } x \notin C\} = \{0, 2\}$ List elements in A that are not in C.

e) $B \cap C = \{0, 2, 4\} \cap \{1, 3, 5\}$
$= \{ \quad \}$ There are no elements in both B *and* C, so the set contains no elements.

Note in Example 3e that the set with no elements can also be written $\varnothing$ and is called the **empty set or null set.**

1.2 THE REAL NUMBERS

Sets of numbers

Although we will most often use the set of real numbers throughout this course, we will also use subsets of the real numbers. The table below contains the sets we will use.

Sets of Numbers	Symbol	Description
Natural Numbers	N	$\{1, 2, 3, \ldots\}$
Whole Numbers	W	$\{0, 1, 2, \ldots\}$
Integers	I	$\{\ldots, -3, -2, -1, 0, 1, 2, 3, \ldots\}$
Rational Numbers	Q	$\{\frac{a}{b} \mid a$ and b are integers, $b \neq 0\}$
Irrational Numbers	H	$\{x \mid x$ is a real number that is not rational$\}$
Real Numbers	$\Re$	$\{x \mid x$ is a point on the real number line$\}$

EXAMPLE 4 List the numbers in the set $\{-4, -\sqrt{3}, -\frac{1}{4}, 0, \frac{1}{2}, 3, \sqrt{11}, \frac{10}{2}\}$ that belong to each of the following sets.

a) Natural numbers

b) Whole numbers

c) Integers

d) Rational numbers

e) Irrational numbers

f) Real numbers

SOLUTION 4

a) 3 and $\frac{10}{2}$ are natural numbers.

b) 0, 3, $\frac{10}{2}$ are whole numbers.

c) –4, 0, 3, $\frac{10}{2}$ are integers.

d) –4, $-\frac{1}{4}$, 0, $\frac{1}{2}$, 3, $\frac{10}{2}$ are rational numbers.

e) $-\sqrt{3}$, $\sqrt{11}$ are irrational numbers.

f) all the numbers listed are real numbers.

Notice the subset relationship demonstrated by this example. That is,

$$N \subseteq W \subseteq I \subseteq Q \subseteq H \subseteq \Re$$

EXAMPLE 5 Place $\subseteq$ or $\not\subseteq$ in each space to make a true statement.

a) $\{1, 2, 3\}$ $\quad W$

b) $\{\frac{1}{2}, 4\}$ $\quad I$

c) I $\quad Q$

d) Q $\quad H$

e) H $\quad R$

SOLUTION 5

a) $\{1, 2, 3\} \subseteq W$

b) $\{\frac{1}{2}, 4\} \not\subseteq I$ — Since $\frac{1}{2}$ is not an integer.

c) $I \subseteq Q$ — Since the integers are a subset of the rational numbers.

d) $Q \not\subseteq H$ — Since the rational numbers are not a subset of the irrational numbers.

e) $H \subseteq \Re$ — Since the irrational numbers are a subset of the real numbers.

We use a real number line to picture the sets of numbers described above. We begin by drawing a line and labeling a convenient point as 0. Mark a point to the right of 0 and label it as 1. The distance between 0 and 1 gives us a way to measure and label other numbers on the number line.

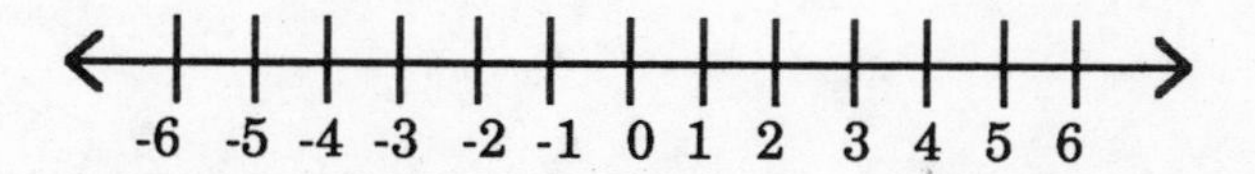

We graph a number on the number line by placing a point at the coordinate of the number. The numbers 4, –2, 0, $1\frac{1}{2}$ are graphed on the number line below.

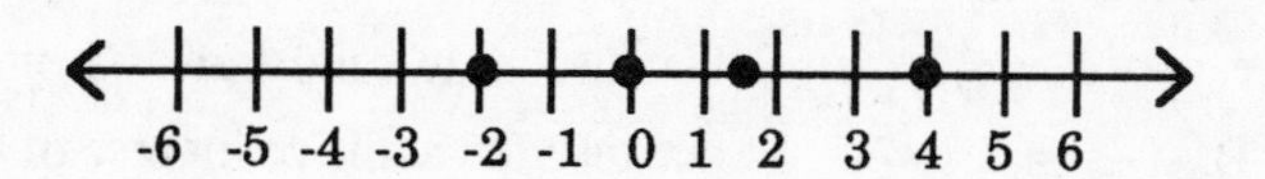

Additive inverse

We can use a number line to define the additive inverse or opposite of a number. The **additive inverse** of a number a is the number that is the same distance from 0 as a, but on the opposite side of 0.

We use the symbol "–" for the additive inverse. For example, the additive inverse of 5 is –5. Note, then, that the additive inverse of –2, written –(–2) is 2, which leads to the following rule:

Additive Inverse:	
$-(-a) = a$	for any number a

Absolute value

We can use a number line to define absolute value. The distance of a number a from zero is its **absolute value,** written $|a|$.

EXAMPLE 6 Find the value of each.

a) $|4|$

b) $|-3|$

c) 0

d) $-|2|$

e) $-|-2|$

f) $-(-2)$

SOLUTION 6

a) $|4| = 4$ because 4 is 4 units away from 0.

b) $|-3| = 3$ because –3 is 3 units away from 0.

c) $0 = 0$ because 0 is 0 units away from 0.

d) $-|2| = -(2)$ because 2 is 2 units away from 0.

$\quad = -2$ because the additive inverse of 2 is –2.

e) $-|-2| = -(2)$ because –2 is 2 units away from 0.

f) $-(-2) = 2$ because the additive inverse of –2 is 2.

Graphing inequalities

We can also use a number line to graph simple and compound inequalities. Simple inequalities consist of one inequality statement such as $x < 2$. Compound inequalities consist of two or more simple inequalities linked by the words "and" or "or."

Simple inequalities

To graph an inequality in the form $x < a$:
1. Put an open circle on a on a number line.
2. Shade to the left of a, using an arrow to indicate that the shading continues.

To graph an inequality in the form $x > a$:
1. Put an open circle on a on a number line.
2. Shade to the right of a, using an arrow to indicate that the shading continues.

If the inequality is ≤ or ≥, use a closed (darkened) circle on a.

EXAMPLE 7 Graph.

a) $\{x | x < -2\}$

b) $\{x | x \geq 3\}$

SOLUTION 7

a) To graph $x < -2$, begin by putting an open circle on -2. Next, shade to the *left* of -2:

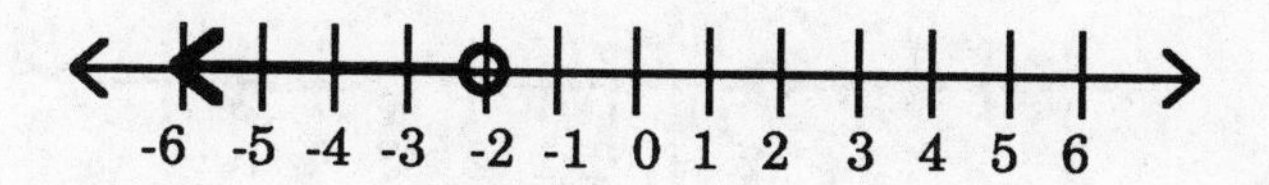

b) To graph $x \geq 3$, begin by putting a closed circle on 3. Next, shade to the *right* of 3:

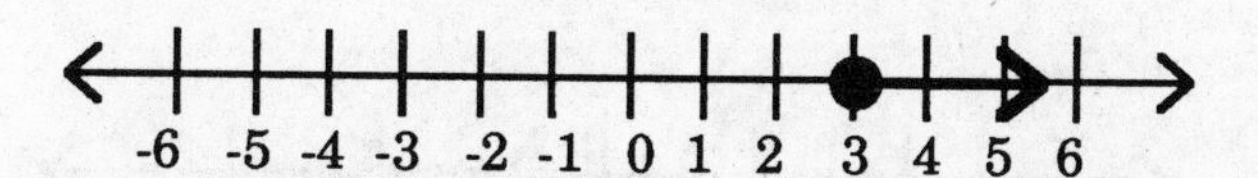

Compound inequalities

We used the union of sets A and B to represent elements found in either set A *or* set B. The words "or" and "and" are also used to form compound inequalities.

EXAMPLE 8 Graph $\{x | x < -2 \text{ or } x \geq 3\}$.

SOLUTION 8

We have already graphed each of the simple inequalities:

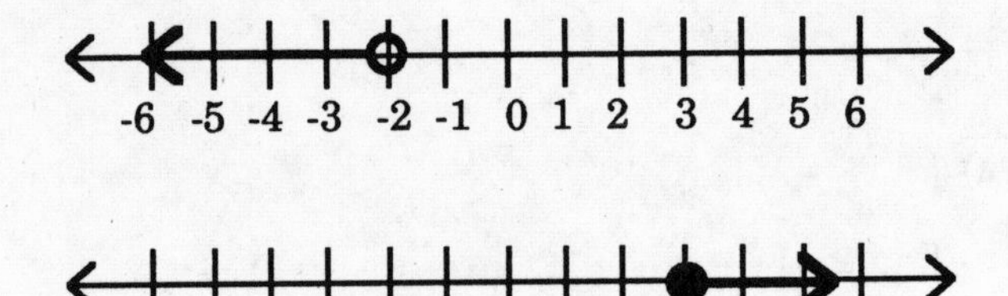

We now join the two graphs using the word "or" so that the final graph contains points from either $x < -2$ or $x \geq 3$:

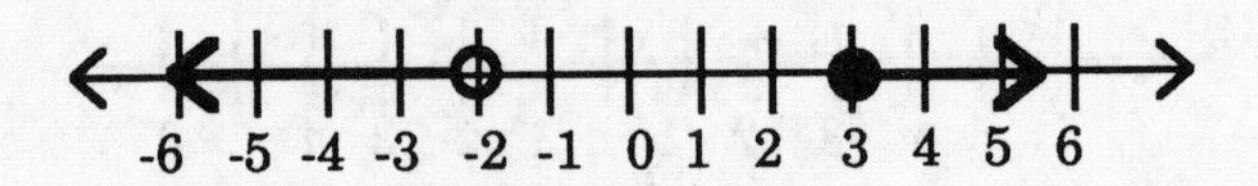

EXAMPLE 9 Graph $\{x | x > 1 \text{ and } x < 4\}$.

SOLUTION 9

Begin by graphing each simple inequality.

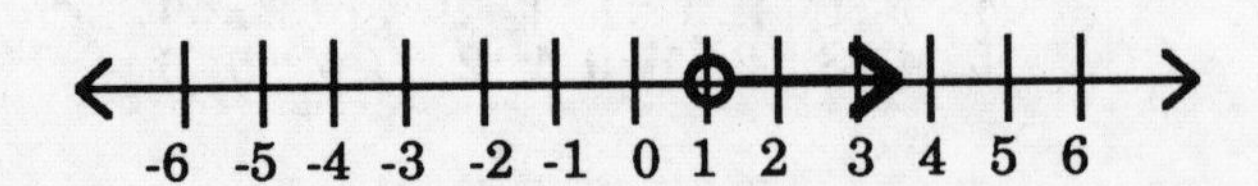

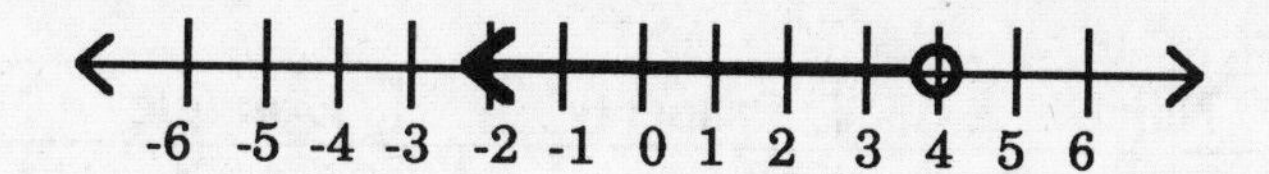

The final graph must contain points from $x > 1$ *and* $x < 4$, that is, the intersection (or overlap) of our simple inequalities.

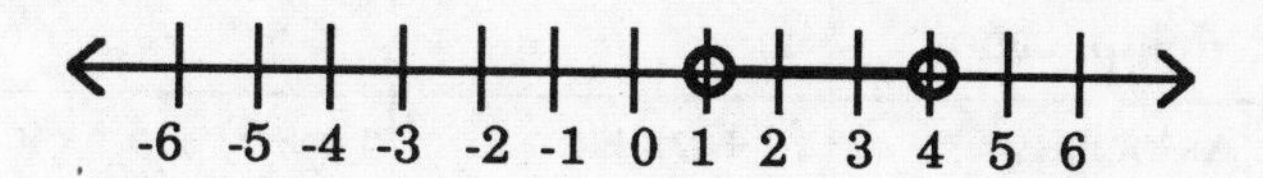

Another way to write $x > 1$ and $x < 4$ is $1 < x < 4$. Note that this shortcut notation can only be used when the compound inequality represents an intersection, using the word "and."

EXAMPLE 10 Graph $\{x| -2 \le x \le 4\}$.

SOLUTION 10

$-2 \le x \le 4$ means $x \ge -2$ *and* $x \le 4$. We need to graph all points between -2 and 4, including -2 and 4:

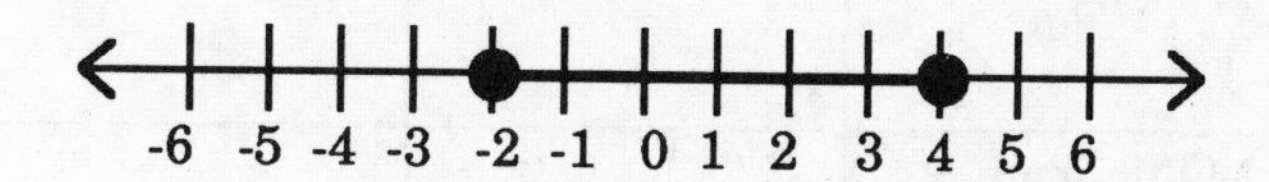

1.3 PROPERTIES OF REAL NUMBERS

The properties of real numbers are rules that hold true for *all* real numbers (rational and irrational numbers). The following is a list of these properties, an example of each property, and what to look for.

Name	Property	Example	Look For
Commutative Property of Addition	$a + b = b + a$	$3 + 4 = 4 + 3$	Change in the order
Commutative Property of Multiplication	$a \cdot b = b \cdot a$	$4 \cdot 6 = 6 \cdot 4$	Change in the order
Associative Property of Addition	$(a + b) + c = a + (b + c)$	$(2 + 3) + 5 = 2 + (3 + 5)$	Grouping symbols around a different pair of numbers
Associative Property of Multiplication	$(a \cdot b) \cdot c = a \cdot (b \cdot c)$	$(2 \cdot 3) \cdot 6 = 2 \cdot (3 \cdot 6)$	Grouping symbols around a different pair of numbers
Additive Identity Property	$a + 0 = a$ and $0 + a = a$	$5 + 0 = 5$ and $0 + 5 = 5$	Adding 0 to a number
Multiplicative Identity Property	$a(1) = a$ and $1(a) = a$	$6(1) = 6$ and $1(6) = 6$	Multiplying 1 times a number
Additive Inverse Property	$a + (-a) = 0$	$3 + (-3) = 0$	Adding a number and its additive inverse
Multiplicative Inverse Property	$a(\frac{1}{a}) = 1$	$6(\frac{1}{6}) = 1$	Multiplying a number times 1 over the number
Distributive Property	$a(b + c) = a(b) + a(c)$ and $(b + c)a = b(a) + c(a)$	$2(3 + x) = 2(3) + 2(x)$	Multiplying across parentheses

EXAMPLE 11 State the property that justifies each statement.

a) $-2 + 5 = 5 + (-2)$

b) $4(3x) = (4 \cdot 3)x$

c) $8 + 0 = 8$

d) $3\left(\frac{1}{3}\right) = 1$

SOLUTION 11

a) $-2 + 5 = 5 + (-2)$ Commutative Property of Addition. The order of the numbers changed, and addition was the operation.

b) $4(3x) = (4 \cdot 3)x$ Associative Property of Multiplication. The order of the factors 4, 3, x stayedthe same on both sides of the equal sign. The parentheses first grouped 3 and x, then 4 and 3.

c) $8 + 0 = 8$ Additive Identity Property. Zero added to any number equals the original number.

d) $3\left(\frac{1}{3}\right) = 1$ Multiplicative Inverse Property. 3 and $\frac{1}{3}$ are called multiplicativeinverses, and their product is 1.

EXAMPLE 12 Apply the Distributive Property and simplify the result.

a) $4(3x + 7)$

b) $\frac{1}{2}(4x + 9)$

c) $2(3a + 5) + 6$

SOLUTION 12

a) $4(3x + 7) = 4(3x) + 4(7)$ Apply the Distributive Property.

$= 12x + 28$ Use the Associative Property: $4(3x) = (4 \cdot 3)x = 12x$.

b) $\frac{1}{2}(4x + 9) = \frac{1}{2}(4x) + \frac{1}{2}(9)$ Apply the Distributive Property.

$= 2x + \frac{9}{2}$ $\frac{1}{2}(9) = \frac{1}{2} \cdot \frac{9}{1} = \frac{9}{2}$.

c) $2(3a + 5) + 6 = 2(3a) + 2(5) + 6$ — Apply the Distributive Property.

$= 6a + 10 + 6$ — Use the Associative Property: $2(3a) = (2 \cdot 3)a = 6a$

$= 6a + 16$ — Add.

EXAMPLE 13 Use the Distributive Property to combine similar terms.

a) $6x + 8x$

b) $5y + y$

SOLUTION 13

a) $6x + 8x = (6 + 8)x$ — Apply the Distributive Property.

$= 14x$ — Add.

b) $5y + y = 5y + 1y$ — $y = 1 \cdot y$

$= (5 + 1)y$ — Apply the Distributive Property.

$= 6y$ — Add.

1.4 OPERATIONS WITH REAL NUMBERS

This section contains a review of the four basic operations-addition, subtraction, multiplication and division-with real numbers.

Adding real numbers

You may recall that adding signed numbers in a basic algebra course is first presented with a number line. The number line examples led to the following rules.

Adding Real Numbers
1. To add two numbers with the same sign, add their absolute values and write the common sign.
2. To add two numbers with different signs, subtract their absolute values and use the sign of the number with the larger absolute value.

EXAMPLE 14 Find the sums.

a) $-8 + (-4)$

b) $5 + (-3)$

c) $2 + (-7)$

SOLUTION 14

a) $-8 + (-4) = -(8 + 4)$	Add the absolute values.
$= -12$	Use the common sign, –.
b) $5 + (-3) = 2$	Subtract, and use the sign of 5.
c) $2 + (-7) = -5$	Subtract, and use the sign of –7.

Subtracting real numbers

Subtraction is defined as adding the additive inverse. For example, $6 - 2$ means $6 + (-2)$. This leads to the following rules for subtraction:

Subtracting Signed Numbers
1. Change the subtraction symbol to addition.
2. Write the additive inverse of the number being subtracted.
3. Use the rules for adding signed numbers.

EXAMPLE 15 Subtract.

a) $5 - 9$

b) $-4 - 7$

c) $2 - (-8)$

d) $-3 - (-6)$

SOLUTION 15

a) $5 - 9 = 5 + (-9)$ — Change subtraction symbol to +; additive inverse of 9 is –9.

$= -4$ — Different signs: subtract, use sign of –9.

b) $-4 - 7 = -4 + (-7)$ — Change subtraction symbol to +; additive inverse of 7 is –7.

$= -11$ — Same signs: add, use common sign.

c) $2 - (-8) = 2 + (+8)$ — Change subtraction symbol to + additive inverse of –8 is +8.

$= 10$ — Same signs: add, use common sign.

d) $-3 - (-6) = -3 + (+6)$ — Change subtraction symbol to +; additive inverse of –6 is +6.

$= 3$ — Different signs: subtract, use sign of +6.

Adding and subtracting fractions

The rules for adding and subtracting signed numbers apply to *all* real numbers. In particular, we can use the rules for rational numbers. Recall the basic rules for adding and subtracting fractions:

$\frac{a}{b} + \frac{c}{b} = \frac{a+c}{b}$ Adding like fractions

$\frac{a}{b} - \frac{c}{b} = \frac{a-c}{b}$ Subtracting like fractions

Note from these rules that fractions must have a common denominator before they can be added or subtracted. Therefore, you must rewrite fractions as equivalent fractions with a common denominator before performing additions and subtractions.

EXAMPLE 16 Add or subtract as indicated.

a) $-\frac{5}{6} + (-\frac{2}{9})$

b) $\frac{1}{12} - \frac{7}{8}$

SOLUTION 16

a) $-\frac{5}{6} + (-\frac{2}{9}) = -\frac{15}{18} + (-\frac{4}{18})$

Rewrite each fraction with the common denominator of 18.

$= -\frac{19}{18}$

Same signs: add the numerators, keep the common denominator. Use the common sign.

b) $\frac{1}{2} - \frac{7}{8} = \frac{2}{24} - \frac{21}{24}$

Rewrite each fraction with the common denominator of 24.

Change subtraction symbol to +; additive inverse of $\frac{21}{24}$ is $-\frac{21}{24}$.

$= \frac{2}{24} + (-\frac{21}{24})$

Different signs: subtract the numerators, keep the common denominator. Use the sign of $-\frac{21}{24}$.

$= -\frac{19}{24}$

Multiplying real numbers

In the equation $2 \cdot 8 = 16$, 2 and 8 are factors of 16 and 16 is called the product. We have the following rules for finding the product of two real numbers:

Multiplying Two Real Numbers

1. If both factors have the same sign, the product is positive.
2. If the factors have different signs, the product is negative.

EXAMPLE 17 Find each product.

a) $5(4)$

b) $(-3)(-2)$

c) $2(-7)$

d) $(-4)(6)$

SOLUTION 17

a) $5(4) = 20$ Same signs: positive product.

b) $(-3)(-2) = 6$ Same signs: positive product.

c) $2(-7) = -14$ Different signs: negative product.

d) $(-4)(6) = -24$ Different signs: negative product.

Multiplying fractions

The basic rule for multiplying fractions is

$$\frac{a}{b} \cdot \frac{c}{d} = \frac{ac}{bd}$$

We can combine this rule with our sign rule for multiplication to multiply positive and negative fractions.

EXAMPLE 18 Find each product.

a) $(-\frac{2}{3})(-\frac{3}{5})$

b) $\frac{2}{7}(-7)$

c) $-\frac{5}{12}(\frac{8}{11})$

SOLUTION 18

a) $(-\frac{2}{3})(-\frac{3}{5}) = \frac{2 \cdot 3}{3 \cdot 5}$ Same signs: positive product.

$= \frac{2}{5}$ Reduce: $\frac{3}{3} = 1$.

b) $\frac{2}{7}(-7) = \frac{2}{7}(-\frac{7}{1})$ $-7 = -\frac{7}{1}$.

$= -\frac{2 \cdot 7}{7 \cdot 1}$ Different signs: negative product.

$= -2$ Reduce: $\frac{7}{7} = 1$.

c) $-\frac{5}{12}(\frac{8}{11}) = -\frac{5 \cdot 8}{12 \cdot 11}$ Different signs: negative product.

$= -\frac{10}{33}$ Reduce: $\frac{8}{12} = \frac{2}{3}$.

Note from Example 18 a) that the same nonzero value in the numerator and denominator can be reduced to 1.

Dividing real numbers

Since division is defined as multiplication by the reciprocal ($\frac{a}{b} = a \cdot \frac{1}{b}$, $b \neq 0$), the sign rules for division are the same as the sign rules for multiplication.

> **Dividing Two Real Numbers**
> 1. If both numbers have the same sign, the quotient is positive.
> 2. If the numbers have different signs, the quotient is negative.

EXAMPLE 19 Divide.

a) $\frac{32}{-8}$

b) $\frac{-56}{-7}$

c) $-16 \div 2$

d) $\frac{-12}{0}$

SOLUTION 19

a) $\frac{32}{-8} = -4$ — Different signs: negative quotient.

b) $\frac{-56}{-7} = 8$ — Same signs: positive quotient.

c) $-16 \div 2 = -8$ — Different signs: negative quotient.

d) $\frac{-12}{0}$ — cannot be done. Division by 0 is *not* defined.

Dividing fractions

The basic rule for dividing fractions is

$$\frac{a}{b} \div \frac{c}{d} = \frac{a}{b} \cdot \frac{d}{c} = \frac{ad}{bc} \qquad b, d \neq 0$$

We can combine this rule with our sign rule for division to divide positive and negative fractions.

EXAMPLE 20 Divide.

a) $\frac{1}{4} \div (-\frac{3}{8})$

b) $(-\frac{2}{5}) \div (-\frac{4}{25})$

c) $(-\frac{3}{7}) \div (-\frac{3}{7})$

SOLUTION 20

a) $\frac{1}{4} \div (-\frac{3}{8}) = \frac{1}{4} \cdot (-\frac{8}{3})$ Invert and multiply.

$= -\frac{1 \cdot 8}{4 \cdot 3}$ Different signs: negative product.

$= -\frac{2}{3}$ Reduce.

b) $(-\frac{2}{5}) \div (-\frac{4}{25}) = (-\frac{2}{5}) \cdot (-\frac{25}{4})$ Invert and multiply.

$= \frac{2 \cdot 25}{5 \cdot 4}$ Same signs: positive product.

$= \frac{5}{2}$ Reduce.

c) $(-\frac{3}{7}) \div (-\frac{3}{7}) = (-\frac{3}{7}) \cdot (-\frac{7}{3})$ Invert and multiply.

$= \frac{3 \cdot 7}{7 \cdot 3}$ Same signs: positive

product.

$= 1$ Reduce.

Order of operations

We can now combine our rules for order of operations with our rules for operations with real numbers.

EXAMPLE 21 Simplify.

a) $(-3-7)(-4-1)$

b) $4-7(3-5)$

c) $3(-2)^2-3^2$

d) $\dfrac{-4(-6)+8(-2)}{6-2\cdot 5}$

e) $\dfrac{4^3+3^3}{4^2-3^2}$

f) $-2|-5-6|-3|2-5|$

SOLUTION 21

a) $(-3-7)(-4-1) = (-10)(-5)$ — Work inside parentheses first.

$= 50$ — Same signs: positive product.

b) $4-7(3-5) = 4-7(-2)$ — Work inside parentheses first.

$= 4+14$ — Same signs: positive product.

$= 18$ — Add.

c) $3(-2)^2-3^2 = 3(4)-9$ — $(-2)^2 = -2 \cdot -2 = +4$; $3^2 = 3 \cdot 3 = 9$

$= 12-9$ — Multiply before subtracting.

$= 3$ — Subtract.

d) $\dfrac{-4(-6)+8(-2)}{6-(2\cdot 5)} = \dfrac{24+(-16)}{6-10}$ Simplify numerator and denominator.

$= \dfrac{8}{-4}$

$= -2$ Different signs: negative quotient.

e) $\dfrac{4^3+3^3}{4^2-3^2} = \dfrac{64+27}{9-16}$ Exponents first.

$= \dfrac{91}{-7}$ Simplify the numerator and the denominator.

$= -13$ Different signs: negative quotient.

f) $-2|-5-6|-3|2-5| = -2|-11|-3|-3|$ Work inside absolute value bars first.

$= -2(11) - 3(3)$ $|-11| = 11; |-3| = 3.$

$= -22 - 9$ Multiply before subtracting.

$= -31$ Subtract.

Practice Exercises

1. Translate each statement into symbols

(a) The sum of x and 6 is greater than or equal

to

18.

(b) The difference of x and 4 is less than 8.

(c) The product of 8 and y is 24.

(d) The quotient of x and 8 is 6 more than the difference of x and 2.

2. Simplify each expression.

(a) $2+4(1+8)$

(b) $2 \cdot 6^2 - 3 \cdot 4^2$

(c) $4+24 \div 6 \cdot 2$

(d) $6 + 2[18 - 3(5 - 2)]$

3. Let $A = \{1, 3, 5, 7\}$, $B = \{0, 2, 4\}$, $C = \{1, 2, 3, 4\}$.

Find each set.

(a) $B \cup C$

(b) $B \cap C$

(c) $A \cap B$

(d) $\{x | x \in A \text{ and } x \notin C\}$

4. List the numbers in the set $\{-1, -\frac{1}{2}, 0, \sqrt{2}, 2, \frac{6}{2}\}$ that belong to

each

of the following sets.

(a) Natural numbers

(b) Whole numbers

(c) Integers

(d) Rational numbers

(e) Irrational numbers

(f) Real numbers

5. Graph.

(a) $\{x | x \geq 2\}$

(b) $\{x | x > -1\}$

(c) $\{x | x \leq -3 \text{ or } x > 1\}$

(d) $\{x | -2 < x < 4\}$

6. State the property that justifies each statement.

(a) $2 \cdot 6 = 6 \cdot 2$

(b) $5 + (3 + 2) = (5 + 3) + 2$

(c) $1 \cdot x = x$

(d) $5 + (-5) = 0$

(e) $4(2x + 3) = 8x + 12$

7. Apply the distributive property and simplify.

(a) $2(3x + 1)$

(b) $\frac{1}{3}(6x+9)$

(c) $5(2x + 3) + 4$

(d) $6(3x + 2) + 5$

8. Perform the indicated operations.

(a) $-3 + (-6)$

(b) $15 + (-8)$

(c) $3 + (-12)$

(d) $8 - 14$

(e) $-3 - 17$

(f) $4 - (-6)$

9. Perform the indicated operations.

(a) $6(-2)$

(b) $(-4)(-3)$

(c) -3(5)

(d) 0(-8)

10. Perform the indicated operations.

(a) $-\frac{4}{15} + (-\frac{3}{10})$

(b) $\frac{7}{12} - \frac{5}{8}$

(c) $-\frac{5}{6} - (-\frac{2}{9})$

(d) $(-\frac{4}{6})(-\frac{14}{6})$

(e) $\frac{2}{9}(-\frac{9}{14})$

(f) $(-\frac{5}{8}) \div (-\frac{10}{24})$

11. Simplify.

(a) (6 - (-2))(-5 - 7)

(b) 3 - 6(2 - 7)

(c) $-4^2 - 6^2$

(d) $\frac{5(-3) - 4(-6)}{-2 - 4}$

(e) $\frac{2^3 - 3^3}{2^2 - 3^2}$

(f) $-|-3-4| + |-2+6|$

Answers

1.

(a) $x + 6 \geq 18$

(b) $x - 4 < 8$

(c) $8y = 24$

(d) $\frac{x}{8} = 6 + (x - 2)$

2.

(a) 38

(b) 24

(c) 12

(d) 24

3.

(a) $\{0, 1, 2, 3, 4\}$

(b) $\{2, 4\}$

(c) $\varnothing$

(d) $\{5, 7\}$

4.

(a) $2, \frac{6}{2}$

(b) $0, 2, \frac{6}{2}$

(c) $-1, 0, 2, \frac{6}{2}$

(d) $-1, -\frac{1}{2}, 0, 2, \frac{6}{2}$

(e) $\sqrt{2}$

(f) All numbers listed are real numbers

5.(a)

-6 -5 -4 -3 -2 -1 0 1 2 3 4 5 6

(b)

(c)

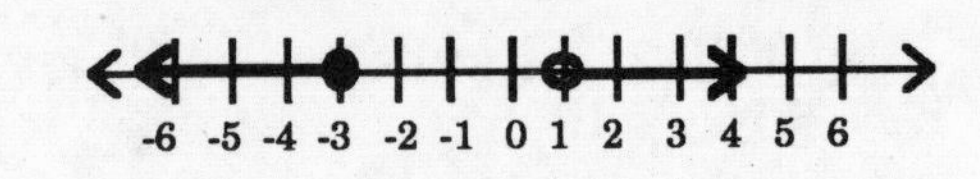

(d)

6.

(a) Commutative Property of Multiplication

(b) Associative Property of Addition

(c) Multiplicative Identity Property

(d) Additive Inverse Property

(e) Distributive Property

7.

(a) $6x + 2$

(b) $2x + 3$

(c) $10x + 19$

(d) $18x + 17$

8.

(a) -9

(b) 7

(c) -9

(d) -6

(e) -20

(f) 10

9.

(a) -12

(b) 12

(c) -15

(d) 0

10.

(a) $-\frac{17}{30}$

(b) $-\frac{1}{24}$

(c) $-\frac{11}{18}$

(d) $\frac{4}{3}$

(e) $-\frac{1}{7}$

(f) $\frac{3}{2}$

11.

(a) -96

(b) 33

(c) -52

(d) $-\frac{3}{2}$

(e) $\frac{19}{5}$

(f) -3

2

Solving Linear Equations and Inequalities

2.1 SOLVING LINEAR EQUATIONS IN ONE VARIABLE

A **linear equation** in one variable can be written as

$$ax + b = c$$

where a, b, and c are real numbers and $a \neq 0$. A **solution** to an equation is a number that replaces the variable and makes both sides of the equation equal. For example, $x - 8 = 10$ has 18 as the solution since $18 - 8 = 10$. The **solution set** for an equation is the set containing the solution(s) to the equation. The solution set for $x - 8 = 10$ is $\{18\}$.

Simplifying expressions

One of the first steps we'll use to solve equations is to simplify each side of the equation. In the expression $4x + 8x + 5$, $4x$, $8x$, and 5 are called *terms*. A term is a number or product of a number and variable(s) raised to powers. When terms have exactly the same variables and exponents, they are called *similar terms* or *like terms*. The $4x$ and $8x$ are similar terms. The 4 in $4x$ and 8 in $8x$ are called *numerical coefficients*, the number part of a term. Similar terms can be added or subtracted by adding or subtracting their numerical coefficients (by using the distributive property).

EXAMPLE 1 Simplify each expression.

a) $4x + 8x + 5$

b) $6(y + 1) - 8y$

c) $-2(a + 4) + 5a - 3$

d) $\frac{1}{2}(4x + 6) + 3x$

SOLUTION 1

a) $4x + 8x + 5 = (4 + 8)x + 5$ — Use the distributive property.

$= 12x + 5$ — $12x$ and 5 are not similar terms and so cannot be added.

b) $6(y + 1) - 8y = 6y + 6 - 8y$ — Use the distributive property.

$= -2y + 6$ — Combine numerical coefficients of $6y$ and $-8y$.

c) $-2(a + 4) + 5a - 3 = -2a - 8 + 5a - 3$ — Use the distributive property.

$= 3a - 11$ — Combine similar terms.

d) $\frac{1}{2}(4x + 6) + 3x = \frac{1}{2}(4x) + \frac{1}{2}(6) + 3x$ — Use the distributive property.

$= 2x + 3 + 3x$ — Multiply.

$= 5x + 3$ — Combine similar terms.

Addition and multiplication property of equality

Once both sides of an equation are simplified, the solution is found by using the Addition Property of Equality and the Multiplication Property of Equality. Recall that equivalent equations are equations with the same solutions.

> **Addition Property of Equality**
> You may add the same quantity to both sides of an equation to produce an equivalent equation.

> **Multiplication Property of Equality**
> You may multiply both sides of an equation by the same nonzero quantity to produce an equivalent equation.

Because subtraction is defined as adding the additive inverse, we can use the addition property to subtract the same quantity from both sides of an equation. Also, since division is defined as multiplication by the reciprocal, we can use the multiplication property to divide both sides of an equation by the same nonzero quantity.

EXAMPLE 2 Find the solution set for each equation. Check your solutions.

a) $4x - 8 = -20$

b) $\frac{1}{4}x + 6 = 3$

c) $2x - 3 = 9 - x$

SOLUTION 2

a) $4x - 8 = -20$

$4x - 8 + 8 = -20 + 8$	Add 8 to both sides.
$4x = -12$	
$\frac{4x}{4} = -\frac{12}{4}$	Divide both sides by 4.
$x = -3$	Proposed solution.
Check:	
$4x - 8 = -20$	Original equation.
$4(-3) - 8 \stackrel{?}{=} -20$	Replace the variable with its proposed solution.
$-12 - 8 \stackrel{?}{=} -20$	Simplify the left side.
$-20 = -20$	Both sides are equal, so the solution checks.

The solution set is $\{-3\}$.

b) $\frac{1}{4}x + 6 = 3$

$\frac{1}{4}x + 6 - 6 = 3 - 6$ Subtract 6 from both sides.

$\frac{1}{4}x = -3$

$4 \cdot \frac{1}{4}x = 4(-3)$ Multiply both sides by 4.

$x = -12$ Proposed solution.

Check: $\frac{1}{4}x + 6 = 3$ Original equation.

$\frac{1}{4}(-12) + 6 \stackrel{?}{=} 3$ Replace the variable with its proposed solution.

$-3 + 6 \stackrel{?}{=} 3$ Simplify the left side.

$3 = 3$ Both sides are equal as the solution checks.

The solution set is $\{-12\}$.

c) $2x - 3 = 9 - x$

$2x - 3 + x = 9 - x + x$ Add x to both sides.

$3x - 3 = 9$

$3x - 3 + 3 = 9 + 3$ Add 3 to both sides.

$3x = 12$

$\frac{3x}{3} = \frac{12}{3}$ Divide both sides by 3.

$x = 4$ Proposed solution.

Check: $2x - 3 = 9 - x$ Original equation.

$2(4) - 3 \stackrel{?}{=} 9 - (4)$ Replace the variable with its proposed solution.

$8 - 3 \stackrel{?}{=} 9 - 4$ Simplify each side.

$5 = 5$ Both sides are equal, so the solution checks.

The solution set is $\{4\}$.

Solving linear equations

Example 2 required the use of the addition and multiplication properties of equality. When the left or right side of the equation can be simplified, that simplification is done first. Putting our rules together provides steps for solving linear equations.

Solving a Linear Equation

1. Simplify the left side of the equation by removing parentheses and adding and subtracting similar terms.
2. Simplify the right side of the equation by removing parentheses and adding and subtracting similar terms.
3. Use the Addition Property of Equality to get the variable terms together on one side of the equation and the numbers on the other side of the equation.
4. Multiply both sides by the reciprocal of the numerical coefficient of the variable (or divide both sides by the numerical coefficient of the variable).

EXAMPLE 3 Find the solution set for each equation.

a) $25 + 10(12 - x) = 5(2x - 7)$

b) $\frac{a}{4} - \frac{2a}{3} = 10$

c) $-0.05x - 0.04 = -0.06(x + 3) + 0.1$

SOLUTION 3

a) $25 + 10(12 - x) = 5(2x - 7)$

$25 + 120 - 10x = 10x - 35$ Use the distributive property

$145 - 10x = 10x - 35$ Add similar terms.

$145 - 10x - 10x = 10x - 35 - 10x$ Subtract $10x$ from both sides.

$145 - 20x = -35$

$145 - 20x - 145 = -35 - 145$ Subtract 145 from both sides.

$-20x = -180$

$$\frac{-20x}{-20} = \frac{-180}{-20}$$ Divide both sides by –20.

$$x = 9$$

Check that the solution set is {9} by substituting 9 for x in the original equation.

b) $\frac{a}{4} - \frac{2a}{3} = 10$ Fractions can be eliminated by first multiplying both sides of the equation by 12 (the least common denominator.)

$$12\left(\frac{a}{4} - \frac{2a}{3}\right) = 12(10)$$ Multiply both sides by the LCD.

$$12\left(\frac{a}{4}\right) - 12\left(\frac{2a}{3}\right) = 120$$ Use the distributive property.

$$3a - 8a = 120$$

$$-5a = 120$$ Combine similar terms.

$$\frac{-5a}{-5} = \frac{120}{-5}$$ Divide both sides by –5.

$$a = -24$$

Check that the solution set is {–24} by substituting –24 for a in the original equation.

c) $-0.05x - 0.04 = -0.06(x + 3) + 0.1$

An equation containing decimals can be simplified by first multiplying both sides by a power of 10 to eliminate the decimals. Here we'll multiply both sides of the equation by 100.

$$100(-0.05x - 0.04) = 100[-0.06(x + 3) + 0.1]$$

$$-5x - 4 = -6(x + 3) + 10$$ Use the distributive property.

$$-5x - 4 = -6x - 18 + 10$$ Use the distributive property on the right side.

$$-5x - 4 = -6x - 8$$ Simplify the right side.

$$-5x - 4 + 6x = -6x - 8 + 6x$$ Add $6x$ to both sides.

$$x - 4 = -8$$

$x - 4 + 4 = -8 + 4$ Add 4 to both sides.

$x = -4$

Check that the solution set is $\{-4\}$ by substituting -4 for x in the original equation.

Types of solution sets

The previous examples had solution sets with exactly one solution. There are equations whose solution sets are empty, written $\varnothing$, and others whose solution sets contain all real numbers, written $\Re$.

EXAMPLE 4 Find the solution set.

a) $4(x - 2) = 7 + 4x$

b) $\frac{1}{2}(6x + 4) = x + 2(x + 1)$

SOLUTION 4

a) $4(x - 2) = 7 + 4x$

$4x - 8 = 7 + 4x$ Use the distributive property.

$4x - 8 - 4x = 7 + 4x - 4x$ Subtract $4x$ from both sides.

$-8 = 7$

There are no variables, and this is a false statement. This means there are no values for x that will make the original equation true. The solution set is empty, $\varnothing$.

b) $\frac{1}{2}(6x + 4) = x + 2(x + 1)$

$3x + 2 = x + 2x + 2$ Use the distributive property.

$3x + 2 = 3x + 2$ Simplify the right side.

$3x + 2 - 3x = 3x + 2 - 3x$ Subtract $3x$ from both sides.

$2 = 2$

There are no variables, and this is a true statement. This means that *any* value for x will make the original equation true. The solution set is all real numbers, $\Re$.

2.2 FORMULAS

Formulas are equations that involve more than one variable. You may recall

$A = LW$ Area of a rectangle equals Length times Width

$P = 2L + 2W$ Perimeter of a rectangle equals 2 times the Length added to 2 times the Width.

In this section we will evaluate formulas given sufficient information, solve formulas for a specified variable, and use formulas to solve word problems.

Evaluating formulas

If we are given values for all but one variable in a formula, we can substitute the given values into the formula and solve for the remaining variable. Always substitute given values in parentheses to prevent sign or order of operations errors.

EXAMPLE 5 Find b in the formula $y = mx + b$ when $y = 4$, $m = -2$, and $x = -3$.

SOLUTION 5

$y = mx + b$	Given formula.
$(4) = (-2)(-3) + 6$	Substitute the given values, using parentheses.
$4 = 6 + b$	Simplify the right side of the equation.
$4 - 6 = 6 + b - 6$	Subtract 6 from both sides.
$-2 = b$	

If you had written $-2 - 3$ for mx instead of $(-2)(-3)$, you may have (mistakenly) simplified to -5 instead of $+6$.

EXAMPLE 6 The formula for simple interest is $I = Prt$. Find the principal, P, if \$800 interest was earned at a rate of 10% for 4 years.

SOLUTION 6

In this formula, the rate must be converted to a decimal: $10\% = 0.10$.

The length of time, t, is the 4 years.

$I = Prt$ — Given formula.

$(800) = P(0.10)(4)$ — Substitute the given values, using parentheses.

$800 = P(0.4)$ — Simplify the right side.

$\frac{800}{0.4} = \frac{P(0.4)}{0.4}$ — Divide both sides by 0.4.

$2000 = P$

The principal was \$2000.

Solving formulas for a variable

The last example with $I = Prt$ is solved for I, that is, I is alone on one side of the equation. If you had several problems to find the principal, P, it would be easier to first solve the formula for P. Follow the rules for solving a linear equation when you need to solve a formula for a specified variable.

EXAMPLE 7 Solve $I = Prt$ for P

SOLUTION 7

$I = Prt$ — Given formula.

$\frac{I}{rt} = \frac{Prt}{rt}$ — The coefficient of P is rt. Divide both sides by rt.

$\frac{I}{rt} = P$ — The formula is now solved for P.

EXAMPLE 8 Solve each formula for the specified variable.

a) $y = mx + b$ for x

b) $A = \frac{1}{2}bh$ for b

c) $S = 2\pi rh + 2\pi r^2$ for h

SOLUTION 8

a) $y = mx + b$ — Given formula.

$y - b = mx + b - b$ — Subtract b from both sides.

$y - b = mx$

$$\frac{y-b}{m} = \frac{mx}{m}$$

The coefficient of x is m. Divide both sides by m.

$$\frac{y-b}{m} = x$$

The formula is now solved for x.

b) $A = \frac{1}{2}bh$

Given formula.

$$2(A) = 2\left(\frac{1}{2}bh\right)$$

Multiply both sides by 2 to eliminate the fraction.

$$2A = bh$$

$$\frac{2A}{h} = \frac{bh}{h}$$

Divide both sides by h.

$$\frac{2A}{h} = b$$

The formula is now solved for b.

c) $S = 2\pi rh + 2\pi r^2$

Given formula.

$$S - 2\pi r^2 = 2\pi rh + 2\pi r^2 - 2\pi r^2$$

Subtract $2\pi r^2$ from both sides.

$$S - 2\pi r^2 = 2\pi rh$$

$$\frac{S - 2\pi r^2}{2\pi r} = \frac{2\pi rh}{2\pi r}$$

Divide both sides by $2\pi r$.

$$\frac{S - 2\pi r^2}{2\pi r} = h$$

The formula is now solved for h.

Using formulas to solve word problems

Lists of formulas can usually be found in the back of textbooks, or on cards that come with calculators. Notice that the variables are often the first letter of the quantity the variable represents, such as P for principal and I for interest.

EXAMPLE 9 Joan traveled 155 miles in $2\frac{1}{2}$ hours. Find her average speed.

SOLUTION 9

The formula needed is $d = rt$ where d is distance, r is rate, and t is time. To find her speed (or rate), solve for r:

$d = rt$	Write the formula.
$\frac{d}{t} = \frac{rt}{t}$	Divide both sides by t.
$\frac{d}{t} = r$	The formula is now solved for r.
$\frac{155}{2.5} = r$	Substitute $d = 155$, $t = 2\frac{1}{2} = 2.5$.
$62 = r$	

The average speed was 62 miles per hour.

EXAMPLE 10 The thermometer at the bank reads 35°C. Find the Fahrenheit temperature.

SOLUTION 10

The formula that relates Celsius to Fahrenheit is $C = \frac{5}{9}(F - 32)$. Solve first for F.

$C = \frac{5}{9}(F - 32)$	Formula for Celsius.
$9(C) = 9[\frac{5}{9}(F - 32)]$	Multiply both sides by 9 to eliminate the fraction.
$9C = 5(F - 32)$	
$9C = 5F - 160$	Use the distributive property.
$9C + 160 = 5F - 160 + 160$	Add 160 to both sides.
$9C + 160 = 5F$	
$\frac{9C + 160}{5} = \frac{5F}{5}$	Divide both sides by 5.
$\frac{9C + 160}{5} = F$	The formula is now solved for F.

Now solve for F when $C = 35$.

$$\frac{9(35) + 160}{5} = F$$ Substitute $C = 35$.

$$\frac{315 + 160}{5} = F$$ Multiply in the numerator before adding.

$$95 = F$$

Thus the Fahrenheit temperature was 95°.

2.3 SOLVING LINEAR INEQUALITIES IN ONE VARIABLE

A linear inequality is similar to a linear equation, except that the equality symbol (=) is replaced by one of the four inequality symbols ($<, >, \leq, \geq$). The first four steps for solving a linear inequality are identical to the steps we used for solving a linear equation.

To Solve a Linear Inequality

1. Simplify the left side of the inequality by removing parentheses and adding and subtracting similar terms.
2. Simplify the right side of the inequality by removing parentheses and adding and subtracting similar terms.
3. Use addition to get the variable terms together on one side of the inequality and the numbers on the other side of the inequality.
4. Multiply both sides by the reciprocal of the numerical coefficient of the variable (or divide both sides by the numerical coefficient of the variable).
5. *If you multiply (or divide) by a negative number, you must reverse the inequality symbol.*

Step 5 was not necessary for equations, but notice what happens to an inequality when we multiply both sides by a negative number:

$$-3 < 5$$ Given inequality.

$$-2(-3)\ ?\ -2(5)$$ Multiply both sides by -2.

$$6\ ?\ -10$$

Since $6 > -10$, we must reverse the original inequality symbol. So, if you multiply or divide both sides of an inequality by a negative number, make one of the indicated changes:

Multiplying or Dividing by a Negative

Change:	To:
$<$	$>$
$\leq$	$\geq$
$>$	$<$
$\geq$	$\leq$

EXAMPLE 11 Solve and graph.

a) $3x - 17 < 4$

b) $-2m \geq 6$

c) $8a + 15 \geq 2a + 3$

d) $5(2x - 8) - 5 > 5(x - 3)$

SOLUTION 11

a) $3x - 17 < 4$

$3x - 17 + 17 < 4 + 17$ Add 17 to both sides.

$3x < 21$

$\frac{3x}{3} < \frac{21}{3}$ Divide both sides by 3.

$x < 7$

The solution set is written $\{x | x < 7\}$. The graph has an open circle on 7 and shading to the left:

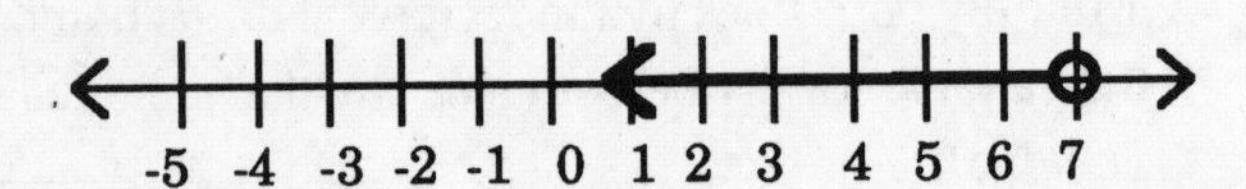

b) $-2m \geq 6$

$\frac{-2m}{-2} \leq \frac{6}{-2}$ Divide both sides by -2. Reverse the inequality symbol.

$m \leq -3$

The solution set is $\{m | m \leq -3\}$. The graph has a closed circle on -3 and shading to the left:

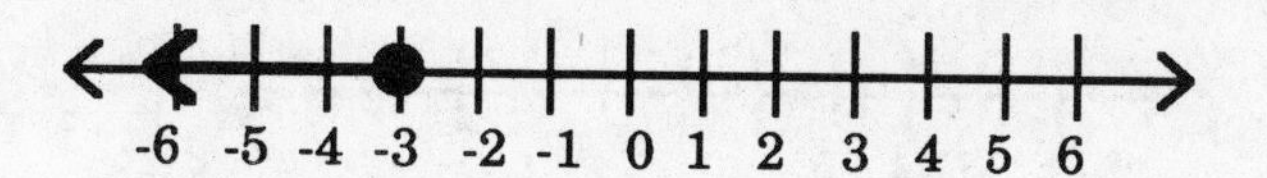

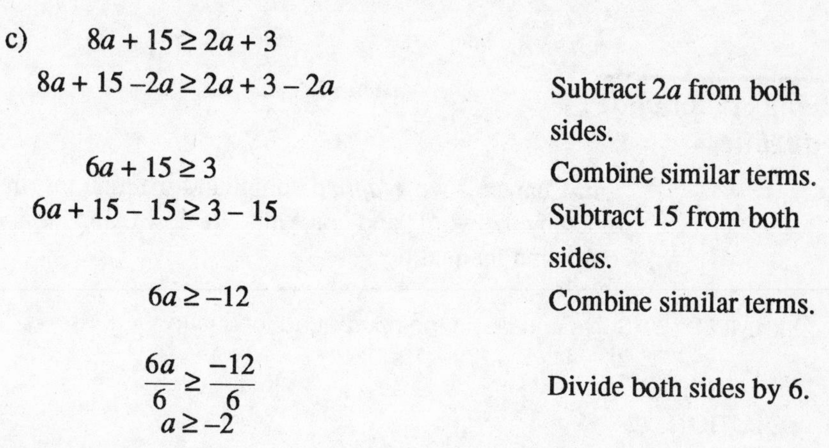

c) $8a + 15 \geq 2a + 3$

$8a + 15 - 2a \geq 2a + 3 - 2a$ Subtract $2a$ from both sides.

$6a + 15 \geq 3$ Combine similar terms.

$6a + 15 - 15 \geq 3 - 15$ Subtract 15 from both sides.

$6a \geq -12$ Combine similar terms.

$\frac{6a}{6} \geq \frac{-12}{6}$ Divide both sides by 6.

$a \geq -2$

The solution set is $\{a | a \geq -2\}$. The graph has a closed circle on -2 and shading to the right.

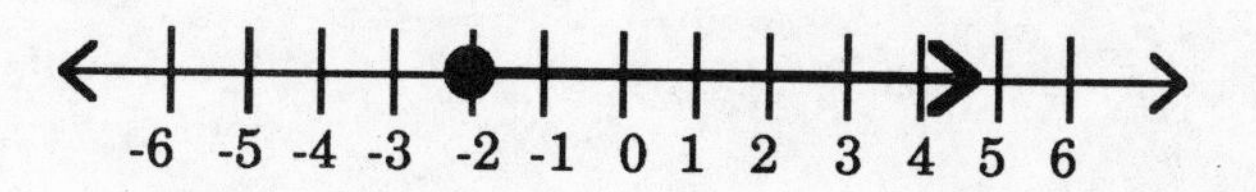

d) $5(2x - 8) - 5 > 5(x - 3)$

$10x - 40 - 5 > 5x - 15$ Use the distributive property to remove parentheses.

$10x - 45 > 5x - 15$ Combine similar terms.

$10x - 45 - 5x > 5x - 15 - 5x$ Subtract $5x$ from both sides.

$5x - 45 > -15$ Combine similar terms.

$5x - 45 + 45 > -15 + 45$ Add 45 to both sides.

$5x > 30$ Add similar terms.

$\frac{5x}{5} > \frac{30}{5}$ Divide both sides by 5.

$x > 6$

The solution set is $\{x | x > 6\}$. The graph has an open circle on 6 and shading to the right.

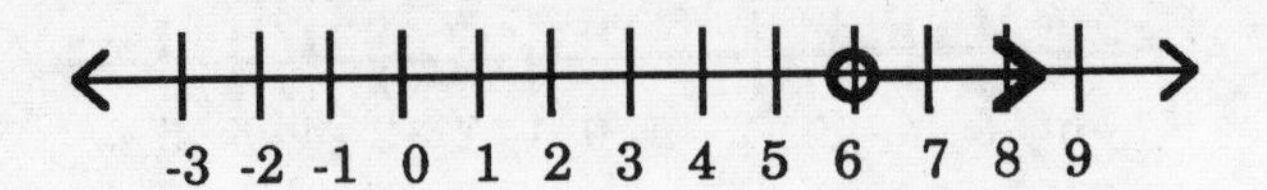

Solving compound inequalities

In Chapter 1 we *graphed* compound inequalities, that is two inequalities joined by the word "and" or "or". We now extend that discussion to *solving* compound inequalities.

EXAMPLE 12 Solve and graph the compound inequality:
$2x - 1 > 7$ or $2x - 1 < -7$

SOLUTION 12

We need to solve each inequality separately:

$2x - 1 > 7$	or	$2x - 1 < -7$	
$2x > 8$		$2x < -6$	Add 1 to both sides.
$x > 4$	or	$x < -3$	Divide both sides by 2.

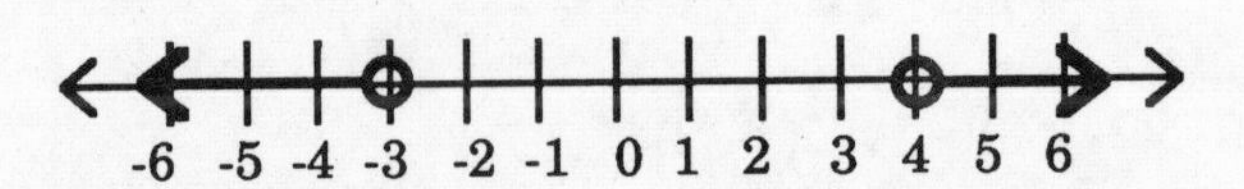

We can write the compound inequality $-7 \le 3x - 2$ and $3x - 2 \le 7$ as $-7 \le 3x - 2 \le 7$. We could solve each inequality separately, but as a shortcut, we'll solve this inequality by working on all three parts at the same time. Keep in mind that x needs to be isolated.

EXAMPLE 13 Solve and graph the compound inequality: $-7 \le 3x - 2 \le 7$

SOLUTION 13

$$-7 \le 3x - 2 \le 7$$

$$-7 + 2 \le 3x - 2 + 2 \le 7 + 2$$ Add 2 to all three parts.

$$-5 \le 3x \le 9$$ Combine similar terms.

$$-\frac{5}{3} \le \frac{3x}{3} \le \frac{9}{3}$$ Divide all three parts by 3.

$-\frac{5}{3} \le x \le 3$

The solution is all real numbers between $-\frac{5}{3}$ and 3, including $-\frac{5}{3}$ and 3.

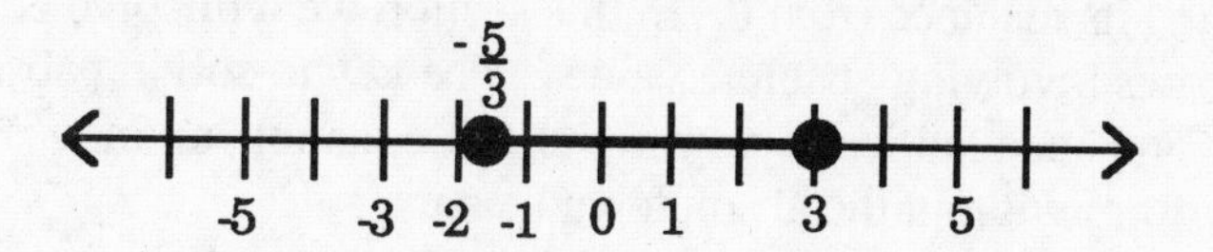

Solving word problems with linear inequalities

Word problems that involve inequalities often use the words *at most, at least, more than, less than.* If x is at most 5, you write $x \le 5$. If x is at least 3, you write $x \ge 3$.

EXAMPLE 14 Robert had test scores of 88, 65, and 75. What must he score on the fourth test to have an average of at least an 80?

SOLUTION 14

Let x = fourth test score then

$$\frac{88 + 65 + 75 + x}{4} \ge 80$$ Use ≥ for at least 80.

$$4\left(\frac{88 + 65 + 75 + x}{4}\right) \ge 4(80)$$ Multiply both sides by 4.

$$88 + 65 + 75 + x \ge 320$$

$$228 + x \ge 320$$ Simplify the left side.

$$228 + x - 228 \ge 320 - 228$$ Subtract 228 from both sides.

$$x \ge 92$$

Thus Robert must score at least 92 on the fourth test.

2.4 SOLVING ABSOLUTE VALUE EQUATIONS AND INEQUALITIES

In Chapter 1 we described the absolute value of a number a, written $|a|$, as its distance from 0. In this section we will solve equations and inequalities involving absolute values. The key to solving absolute value equations and inequalities is to rewrite the equation or inequality in an equivalent form without the absolute value bars.

Absolute Value Form	Equivalent Form Without Absolute Value
$\|x\| = a$	$x = a$ or $x = -a$
$\|x\| < a$	$-a < x < a$
$\|x\| > a$	$x > a$ or $x < a$

Note that $<$ may be replaced by $\leq$, and $>$ may be replaced by $\geq$.

Absolute value equations

The equivalent form for $|x| = 3$ is $x = 3$ or $x = -3$ because both 3 and -3 are 3 units away from 0. Note that this means our solution set has *two* elements, $\{-3, 3\}$.

EXAMPLE 15 Solve.

a) $|x| = 5$

b) $|2x - 1| = 9$

SOLUTION 15

a) $|x| = 5$ — Given absolute value equation.

$x = 5$ or $x = -5$ — Write the equivalent form without absolute value.

The solution set is $\{-5, 5\}$.

b) $|2x - 1| = 9$ — Given absolute value equation.

$2x - 1 = 9$ or $2x - 1 = 9$ — Write the equivalent form without absolute value.

$2x - 1 + 1 = 9 + 1$ $\quad$ $2x - 1 + 1 = -9 + 1$ — Solve each equation.

$2x = 10$ $\quad$ $2x = -8$ — Divide both sides by 2.

$x = 5$ $\quad$ $x = -4$

The solution set is $\{5, -4\}$.

In example 15, the absolute value bars were isolated; that is, the absolute value element must stand alone on one side of the equation. The absolute value bars must be isolated before you can write an equivalent form without the absolute value bars.

EXAMPLE 16 Solve.

a) $|a| - 3 = 10$

b) $|4y - 3| - 5 = 4$

SOLUTION 16

a) $|a| - 3 = 10$

$|a| - 3 + 3 = 10 + 3$ — Isolate the absolute value by adding 3 to both sides.

$|a| = 13$ — Absolute value is isolated.

$a = 13$ or $a = -13$ — Write the equivalent form without absolute value.

The solution set is $\{13, -13\}$.

b) $|4y - 3| - 5 = 4$

$|4y - 3| - 5 + 5 = 4 + 5$ — Isolate the absolute value by adding 5 to both sides.

$|4y - 3| = 9$ — Absolute value is isolated.

$4y - 3 = 9$ or $4y - 3 = -9$ — Write the equivalent form without absolute value.

$4y = 12$ or $4y = -6$ — Solve each equation.

$y = 3$ or $y = -\frac{6}{4}$

$$= -\frac{3}{2}$$

The solution set is $\{3, -\frac{3}{2}\}$.

Solving $|a| = |b|$

If there are two absolute values we can use the following equivalent form:

Absolute Value Form	Equivalent Form
$\lvert a\rvert = \lvert b\rvert$	$a = b$ or $a = -b$

EXAMPLE 17 Solve.

a) $|5x - 4| = |2x - 10|$

b) $|x - 3| = |x - 5|$

SOLUTION 17

a) $|5x - 4| = |2x - 10|$ — Given form $|a| = |b|$.

$5x - 4 = 2x - 10$ or $5x - 4 = -(2x - 10)$ — Write the equivalent form without absolute value.

Solve each equation.

$3x - 4 = -10$	$5x - 4 = -2x + 10$
$3x = -6$	$7x - 4 = 10$
	$7x = 14$
$x = -2$	$x = 2$

b) $|x - 3| = |x - 5|$ — Given form $|a| = |b|$.

$x - 3 = x - 5$ or $x - 3 = -(x - 5)$ — Write the equivalent form without absolute values.

Solve each equation.

$-3 = -5$	$x - 3 = -x + 5$
no variable,	$2x - 3 = 5$
false statement,	$2x = 8$
no solution,	$x = 4$

Only one equation gave us a possible solution. Since 4 does check in the original equation, the solution set is $\{4\}$.

Absolute value inequalities

When the absolute value contains an inequality symbol, choose an equivalent form based on the inequality symbol. These equivalent forms are based on the definition of absolute value as distance from 0. For example, $|x| < 3$ means x must be less than 3 units away from 0:

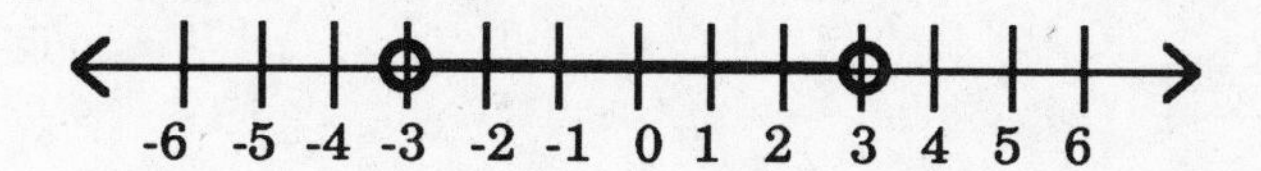

This graph represents $-3 < x < 3$, which is our equivalent form for $|x| < 3$. Similarly, $|x| \geq 1$ means x must be 1 or more units away from 0:

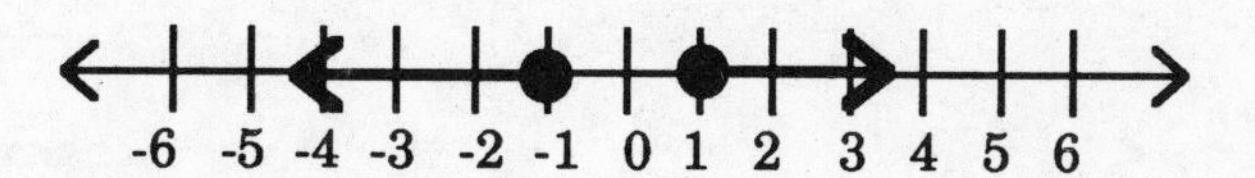

This graph represents $x \leq -1$ or $x \geq 1$, again our equivalent form for $|x| \geq 1$.

EXAMPLE 18 Solve and graph.

a) $|4a - 3| \geq 5$

b) $|3y - 2| \leq 7$

c) $|2x + 3| - 4 > 1$

d) $|3x + 4| - 7 < 3$

SOLUTION 18

a) $|4a - 3| \geq 5$ — Given form of $|x| \geq a$.

$4a - 3 \geq 5$ or $4a - 3 \leq -5$ — Write the equivalent form without absolute value.

$4a \geq 8$ $\quad$ $4a \leq -2$ — Solve each inequality.

$a \geq 2$ $\quad$ $a \leq -\frac{1}{2}$

The solution set is $\{a| \, a \geq 2 \text{ or } a \leq -\frac{1}{2}\}$. The graph has closed circles on 2 and $-\frac{1}{2}$ and shading outward:

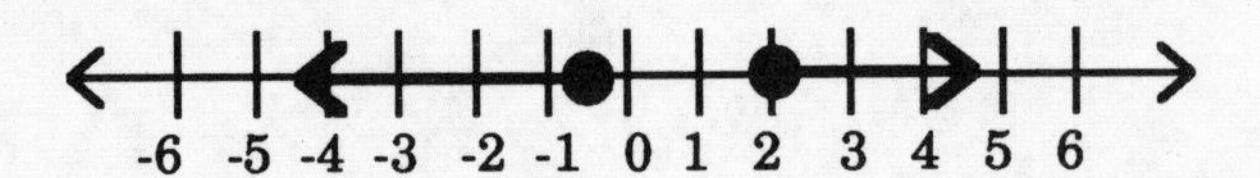

b) $\lvert 3y-2 \rvert \leq 7$	Given form of $\lvert x \rvert \leq a$.
$-7 \leq \;\; 3y-2 \;\; \leq 7$	Write the equivalent form without absolute value.
$-7+2 \leq 3y-2+2 \leq 7+2$	Solve the compound inequality.
$-5 \;\leq\; 3y \;\leq\; 9$	
$-\frac{5}{3} \;\leq\; \frac{3y}{3} \;\leq\; \frac{9}{3}$	Divide by 3.
$-\frac{5}{3} \;\leq\; y \;\leq\; 3$	

The solution set is $\{y| \, -\frac{5}{3} \leq y \leq 3\}$. The graph has closed circles on $-\frac{5}{3}$ and 3, and shading between:

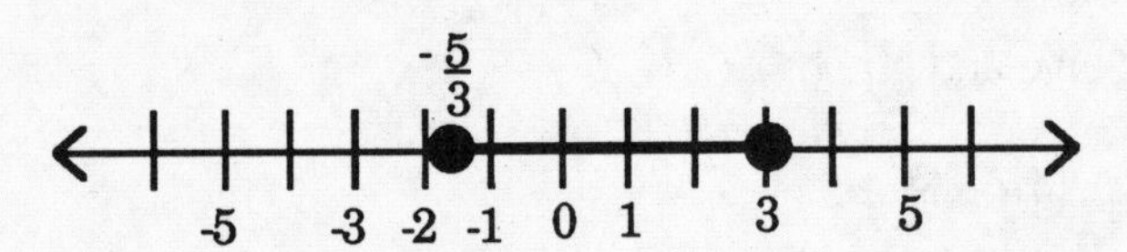

c) $\lvert 2x+3 \rvert - 4 > 1$	First isolate the absolute value by adding 4 to both sides.
$\lvert 2x+3 \rvert > 5$	Absolute value form $\lvert x \rvert > a$.
$2x+3 > 5 \;\text{ or }\; 2x+3 < -5$	Write the equivalent form without absolute value.
$2x > 2 \qquad\qquad 2x < -8$	Solve each inequality.
$x > 1 \;\text{ or }\; x < -4$	

The solution set is $\{x| \, x < 1 \text{ or } x > -4\}$. The graph has open circles on 1 and -4 and shading outward:

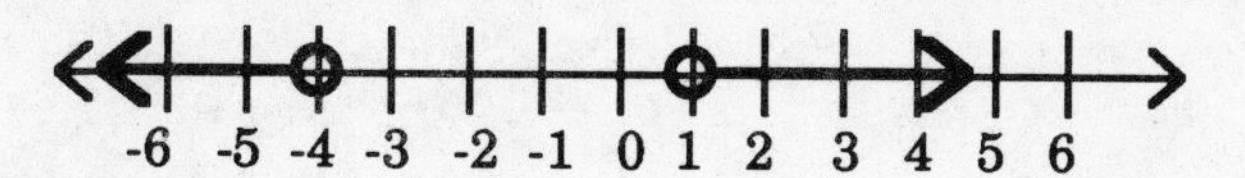

d) $|3x+4|-7<3$ — Isolate the absolute value by adding 7 to both sides.

$|3x+4|<10$ — Absolute value form $|x|<a.$

$-10 \le \quad 3x+4 \quad < 10$ — Write the equivalent form without absolute value.

$-10-4 < 3x+4-4 < 10-4$ — Solve the compound inequality.

$-14 < \quad 3x \quad < 6$

$-\frac{14}{3} < \quad x \quad < \frac{6}{3}$ — Divide by 3.

$-\frac{14}{3} < \quad x \quad < 2$

The solution set is $\{x \mid -\frac{14}{3} < x < 2\}$. The graph has open circles on $-\frac{14}{3}$ and 2 with shading between.

The empty set and all real number solutions

Some special solution sets occur in our work with absolute value equations and inequalities. Because absolute value is defined as a distance, an absolute value cannot be negative. Notice that each of the following examples involve a negative number on the right side of the equation or inequality.

EXAMPLE 19 Solve.

a) $|2x-1| = -4$

b) $|3y-2| < -6$

c) $|5a+3| \geq -1$

d) $|3y-5| - 8 = -4$

SOLUTION 19

a) $|2x-1| = -4$

An absolute value cannot be negative. Thus the solution set is empty, written $\varnothing$.

b) $|3y-2| < -6$

Since an absolute value cannot be negative, surely it cannot be less than a negative number. The solution set is empty, $\varnothing$.

c) $|5a+3| \geq -1$

Every absolute value is larger than a negative number. Thus the solution set is all real numbers, $\Re$.

d) $|3y-5| - 8 = -4$ — Be careful. The absolute value is not isolated yet!

$|3y-5| - 8 + 8 = -4 + 8$ — Add 8 to both sides.

$|3y-5| = 4$ — The right side is not a negative number. You must solve this by writing the equivalent form.

$3y-5 = 4$ or $3y-5 = -4$ — Solve each equation.

$3y = 9$ $\qquad$ $3y = 1$

$y = 3$ or $y = \frac{1}{3}$

The solution set is $\{3, \frac{1}{3}\}$.

2.5 WORD PROBLEMS

Translating words to symbols

Algebra provides a shorthand for symbolizing quantitative relationships. The statement "Cheeseburgers cost 10¢ more than hamburgers" can be symbolized neatly as:

let x = cost of a hamburger
$x + 10$ = cost of a cheeseburger

Translating words to symbols will help you explore and solve word problems. The following table contains some basic symbols and translations:

Addition (+)	**Subtraction** (–)	**Multiplication** (•)	**Division** (÷)	**Equality** (=)
plus	difference	product	quotient	is
increased by	decreased by	times	divided by	is the same as
sum	less than	double (2 times)	reciprocal	is equal to
added to	minus	twice (2 times)	per	are
total		triple (3 times)		equals
more than		factor		

EXAMPLE 20 Translate each phrase into an equation using x as the variable.

a) Twice a number increased by 8 is 18.

b) The product of a number and 5 is 4 less than the number.

c) The sum of a number and double the number is 20 more than the number.

SOLUTION 20

a) $2x + 8 = 18$ — increased by means +

b) $5x = x - 4$ — 4 less than a number means 4 is subtracted *from* the number.

c) $x + 2x = 20 + x$ — 20 more than the number means $20 + x$.

Consecutive integers

One common type of word problem involves consecutive integers, consecutive even integers, or consecutive odd integers. Study these examples:

Term	Using Numbers	Using Variables
Consecutive integers	4, 5, 6, 7	$x, x + 1, x + 2, x + 3$
Consecutive even integers	2, 4, 6, 8	$x, x + 2, x + 4, x + 6$
Consecutive odd integers	3, 5, 7, 9	$x, x + 2, x + 4, x + 6$

EXAMPLE 21 If the sum of the first and third of three consecutive integers is increased by 8, the result is 5 less than triple the second integer. Find the integers.

SOLUTION 21

Let x = first integer

then $x + 1$ = second integer and $x + 2$ = third integer

sum of the first and third	increased by 8	triple the second	5 less than
$x + (x + 2)$	$+ 8$	$= 3(x + 1)$	$- 5$

$x + x + 2 + 8 = 3x + 3 - 5$	Use the distributive property.
$2x + 10 = 3x - 2$	Combine similar terms.
$2x + 10 - 3x = 3x - 2 - 3x$	Subtract $3x$ from both sides.
$-x + 10 = -2$	Combine similar terms.
$-x + 10 - 10 = -2 - 10$	Subtract 10 from both sides.
$-x = -12$	Combine similar terms.
$(-1)(-x) = (-1)(-12)$	Multiply both sides by -1.
$x = 12$	Proposed solution.

Check:

$x + (x + 2) + 8 = 3(x + 1) - 5$	Original equation.
$12 + 14 + 8 \stackrel{?}{=} 3(13) - 5$	Substitute the proposed solutions into the original equation.
$34 = 34$	Both sides are equal, so the solutions check.

Check in the words of the original problem:

The sum of the first and third:	$12 + 14 = 26$
increased by 8:	$+8$
	34
is:	=
5 less than:	-5
triple the second:	$3(13)$
	34

Thus our solution makes sense.

Geometry

EXAMPLE 22 The length of a rectangle is 4 times its width. The perimeter is 60 meters. Find the length and width of the rectangle.

SOLUTION 22

Whenever possible, draw and label a picture.

If we let x = the width of the rectangle

then $4x$ = the length of the rectangle

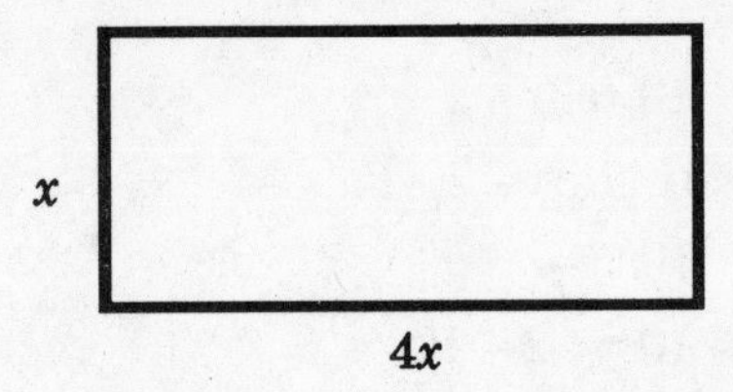

We can use the formula for the perimeter of a rectangle:

$P = 2L + 2W$	Perimeter formula.
$60 = 2(4x) + 2(x)$	Substitute the perimeter, length and width.
$60 = 8x + 2x$	Multiply before adding.
$60 = 10x$	Add similar terms.
$6 = x$	Divide both sides by 10.

Thus the width of the rectangle is 6 meters and the length is $4x = 4(6) = 24$ meters.

Coin problems

An important concept in coin problems involves the value of a number of coins. The value is found by multiplying the number of coins times the value of one coin. For example, 10 nickels have a value of 10(5¢) = 50¢. We'll use this idea to solve the next problem.

EXAMPLE 23 A coin collection consists of nickels and quarters. If there are 17 coins in the collection with a total value of $2.45, how many of each coin are there?

SOLUTION 23

Let x = the number of nickels

then $17 - x$ = the number of quarters

$5(x) + 25(17 - x) = 245$	Multiply the number of coins times the value of one coin to find the total value.
$5x + 425 - 25x = 245$	Use the distributive property.

$-20x + 425 = 245$ Combine similar terms.

$-20x + 425 - 425 = 245 - 425$ Subtract 425 from both sides.

$-20x = -180$ Combine similar terms.

$$\frac{-20x}{-20} = \frac{-180}{-20}$$ Divide both sides by –20.

$$x = 9$$

$$17 - x = 17 - 9 = 8$$

Thus there are 9 nickels and 8 quarters.

Check: 9 nickels = 9(5¢) = 45¢

8 quarters = 8(25¢) = <u>200¢</u>

245¢ = $2.45

so the solutions check.

Interest problems

Interest problems usually involve interest stated as a percent. To find the amount of money earned, we can use the formula $I = Prt$ where the rate is written as a decimal.

EXAMPLE 24 Lisa has $24,000 to invest. She invests part of the money in an account that pays 8% interest, and the remainder in an account that pays 10% interest. If she earns $2080 in interest in one year, how much was invested in each account?

SOLUTION 24

Let x = amount invested at 8%

then $24,000 - x$ = amount invested at 10%

Use $I = Prt$ to find the interest from each account.

Interest at 8% for 1 year	Interest at 10% for 1 year
$I = x(0.08)(1)$	$I = (24000 - x)(0.10)(1)$

The total interest is the sum of the interest from each account:

$$x(0.08)(1) + (24000 - x)(0.10)(1) = 2080$$

$0.08x + 2400 - 0.10x = 2080$ Use the distributive property.

$100[0.08x + 2400 - 0.10x] = 100(2080)$	Multiply both sides by 100 to remove decimals.
$8x + 240000 - 10x = 208000$	Use the distributive property.
$-2x + 240000 = 208000$	Combine similar terms.
$-2x + 240000 - 240000 = 208000 - 240000$	Subtract 240000 from both sides.
$-2x = -32000$	Combine similar terms.
$x = 16000$	Divide both sides by -2.

Lisa has invested \$16,000 at 8% and \$24,000 – \$16,000 = \$8,000 at 10%.

Check:

The interest on \$16,000 at 8% for 1 year = 16,000(0.08)(1) = \$1280

The interest on \$8,000 at 10% for 1 year = 8,000(0.10)(1) = \$800

Total interest = \$1280 + \$800= \$2080

so our solutions check.

Age problems

Age problems can often be organized in a table. Remember that if x represents a person's age today, $x - 5$ represents that person's age 5 years ago, and $x + 2$ represents that person's age 2 years from now.

EXAMPLE 25 Joan is 6 years older than her husband, Bill. Five years ago the sum of their ages was 64. How old are they now?

SOLUTION 25

Let x = Bill's age now

then $x + 6$ = Joan's age now

Let's organize the information in a table:

	Age Now	Age 5 Years Ago
Bill	x	$x - 5$
Joan	$x + 6$	$(x + 6) - 5$

The sum of their ageswas 64

$(x-5)+[(x+6)-5]=64$ Write an equation.

$x-5+x+1=64$ Solve the equation.

$2x-4=64$

$2x=68$

$x=34$

Thus Bill is 34 and Joan is $x+6=34+6=40$.

Check:

If Bill is 34 today, 5 years ago he was 29

If Joan is 40 today, 5 years ago she was <u>35</u>

Total = 64

Thus the solutions check.

Practice Exercises

1. Simplify each expression.

(a) $2y + 8 + 7y$

(b) $3(x + 2) - 9x$

(c) $-4(a + 5) - 3a + 4$

(d) $\frac{2}{3}(6x + 12) + 9x$

2. Find the solution set for each equation.

(a) $4x - 6 = -22$

(b) $\frac{2}{5}x + 8 = 4$

(c) $5x - 17 = 2x + 4$

(d) $25 + 10(12 - x) = 5(2x - 7)$

(e) $\frac{2a}{5} - 7 = \frac{3a}{4}$

(f) $0.6y + 0.3 = 0.5(4y - 2) + 4.9$

3. Find the solution set.

(a) $8 + 3x = 5(x - 4) - 2x$

(b) $\frac{3}{4}(8x + 4) = 3x + 3(x + 1)$

4. Find

(a) m in $y = mx + b$ when $y = 9, x = 5,$ and $b = -6$.

(b) h in $A = \frac{1}{2}bh$ when $A = 12$ and $b = 3$.

(c) b in $A = \frac{h}{2}(a + b)$ when $A = 9, h = 3,$ and $a = 1$.

5. Solve each formula for the specified value.

(a) $y = mx + b$ for b

(b) $A = \frac{1}{2}bh$ for h

(c) $A = \frac{h}{2}(a + b)$ for b

6. Solve and graph.

(a) $5x - 5 \geq 10$ or $5x - 5 \leq -10$

(b) $3 - \frac{3}{4}x > 9$ or $3 - \frac{3}{4}x < -9$

(c) $-7 < 2x - 1 < 7$

(d) $-14 \leq 4 - 3x \leq 14$

7. Solve each word problem.

(a) The product of 5 and a number is greater than the number minus 8. Find all possible values for the number.

(b) The length of a rectangle is 5 inches more than the width. The perimeter cannot exceed 34 inches. Find all possible values of the width.

(c) Cynthia had test scores of 94, 82, 88, and 91. What must she score on the fifth test to have an average of at least 90?

8. Solve.

(a) $|x| + 4 = 7$

(b) $|4y + 2| - 5 = 4$

(c) $|3m + 1| = |2m - 4|$

(d) $|6x - 2| = |3x + 1|$

(e) $|2x - 3| = |2x + 1|$

9. Solve and graph.

(a) $|x - 2| > 4$

(b) $|2x - 1| < 7$

(c) $|4 - 3x| - 4 \leq 11$

(d) $|4x - 3| + 2 \geq 7$

10. Solve.

(a) $|4x + 3| = -2$

(b) $|5a+1| < -4$

(c) $|3y-2| \geq -2$

(d) $|2m-5| - 6 > -10$

11. Translate each phrase into an equation using x as the variable.

(a) Three times a number decreased by 4 is 12.

(b) The sum of twice a number and 6 is 2 more than the number itself.

(c) The product of a number and 5 is the same as 15.

12. Solve.

(a) The sum of three consecutive integers is 14 more than double the largest integer. Find the integers.

(b) If the sum of the first and third of three consecutive odd integers is decreased by 7, the result is 4 more than the second integer. Find the integers.

(c) If the second of three consecutive even integers is multiplied by 3, the result is 24 less than double the sum of the first and third integers. Findthe integers.

13. Solve.

(a) The width of a rectangle is 1 meter more than $\frac{1}{2}$ its length. Find the length and width of the rectangle if its perimeter is 20 meters.

(b) A triangle has a perimeter of 27 inches. The medium side is 3 inches more than the shortest side and the longest side is twice the shortest side. Find the length of each side.

14. Solve.

(a) Joan has been saving dimes and quarters. In two weeks she saved 38 coins with a total value of $6.80. How many of each coin are there?

(b) Pat invested $3800, part at 4% annual interest and the remainder at 6% annual interest. If she earns $200 interest in one year, how much was invested in each account?

(c) Lisa's age exceeds Julie's age by 14 years. Three years ago, Lisa was three times as old as Julie was then. How old are they now?

Answers

1.

(a) $9y + 8$

(b) $-6x + 6$

(c) $-7a - 16$

(d) $13x + 8$

2.

(a) $\{-4\}$

(b) $\{-10\}$

(c) $\{7\}$

(d) $\{9\}$

(e) $\{-20\}$

(f) $\{-4\}$

3.

(a) $\varnothing$

(b) $\Re$

4.

(a) 3

(b) 8

(c) 5

5.

(a) $y - mx$

(b) $\dfrac{2a}{b}$

(c) $\dfrac{2A}{h} - a$

6.(a) $x \geq 3$ or $x \leq -1$

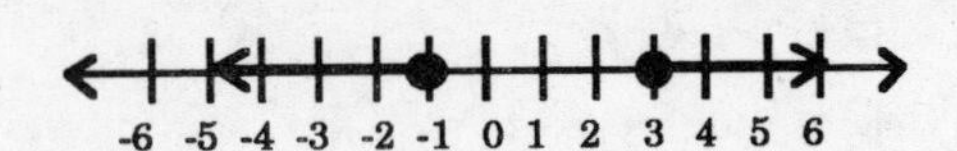

(b) $x < -8$ or $x > 16$

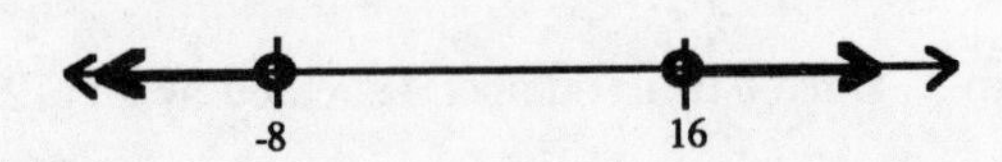

(c) $-3 < x < 4$

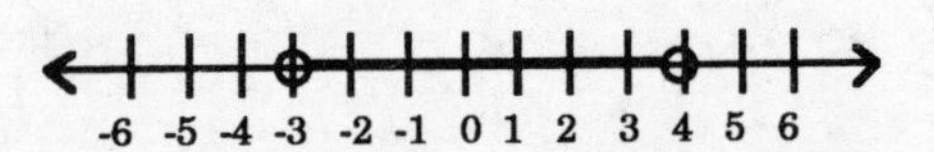

(d) $-\dfrac{10}{3} \leq x \leq 6$

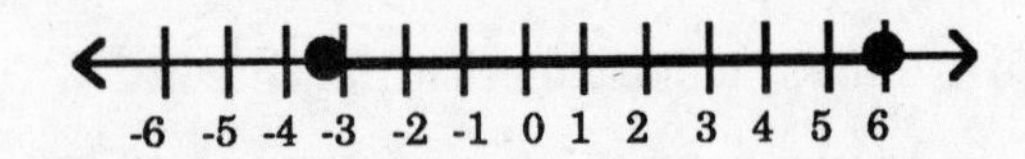

7.

(a) $x > -2$, where x is the number.

(b) $x \leq 2$, where x is the width.

(c) $x \geq 95$, where x is the fifth test score.

8.

(a) $\{-3, 3\}$

(b) $\{-\dfrac{11}{4}, \dfrac{7}{4}\}$

(c) $\{-5, \dfrac{3}{5}\}$

(d) $\{\dfrac{1}{9}, 1\}$

(e) $\{\dfrac{1}{2}\}$

9.

(a) $\{x \mid x > 6 \text{ or } x < -2\}$

(b) $\{x \mid -3 < x < 4\}$

(c) $\{x \mid -\frac{11}{3} \le x \le \frac{19}{3}\}$

(d) $\{x \mid x \ge 2 \text{ or } x \le -\frac{1}{2}\}$

10.

(a) $\varnothing$

(b) $\varnothing$

(c) $\Re$

(d) $\Re$

11.

(a) $3x - 4 = 12$

(b) $2x + 6 = 2 + x$

(c) $5x = 15$

12.

(a) 15, 16, 17

(b) 9, 11, 13

(c) 22, 24, 26

13.

(a) length = 6 meters, width = 4 meters

(b) 6 inches, 9 inches, 12 inches

14.

(a) 20 quarters and 18 dimes

(b) $1400 at 4% and $2400 at 6%

(c) Lisa is 24 and Julie is 10.

3

Exponents and Polynomials

3.1 INTEGER EXPONENTS

Recall from Chapter 1 that we use exponents to write repeated multiplications. For example

$$2^5 = 2 \cdot 2 \cdot 2 \cdot 2 \cdot 2$$

where 2 is called the **base**, and 5 is called the **exponent** or **power**.

EXAMPLE 1 Simplify each expression. State the base and the exponent.

a) 5^2

b) $(-5)^2$

c) -5^2

d) $(\frac{1}{2})^3$

SOLUTION 1

a) $5^2 = 5 \cdot 5 = 25$ Base 5 Exponent 2

b) $(-5)^2 = (-5)(-5) = 25$ Base -5 Exponent 2

c) $-5^2 = -1 \cdot 5 \cdot 5 = -25$ Base 5 Exponent 2

d) $(\frac{1}{2})^3 = \frac{1}{2} \cdot \frac{1}{2} \cdot \frac{1}{2} = \frac{1}{8}$ Base $\frac{1}{2}$ Exponent 3

Be careful with examples like 1b) and 1c). The parentheses in 1b) mean that the base is the entire quantity in parentheses. In 1c), the exponent on 5 is simplified *before* multiplying by –1 because simplifying exponents precedes multiplication in the order of operations.

Zero and negative integer exponents

So far we have used repeated multiplication to simplify exponential expressions. When the exponent is 0, or a negative integer, we need two more definitions.

Definition: $a^0 = 1$ for $a \neq 0$

Definition: $a^{-n} = \frac{1}{a^n}$ for $a \neq 0$

EXAMPLE 2 Simplify the following expressions.

a) 8^0

b) -4^0

c) $6x^0, x \neq 0$

d) 3^{-2}

e) $(-4)^{-2}$

f) $5x^{-3}, x \neq 0$

SOLUTION 2

a) $8^0 = 1$ — Use the definition of 0 as an exponent.

b) $-4^0 = -1 \cdot 4^0$ — Use the order of operations to simplify 4^0 first.

$= -1 \cdot 1$ — Use the definition of 0 as an exponent.

$= -1$

c) $6x^0 = 6 \cdot 1$ Use the definition of 0 as an exponent.

$= 6$

d) $3^{-2} = \frac{1}{3^2}$ Use the definition of a negative exponent.

$= \frac{1}{9}$ $3^2 = 3 \cdot 3 = 9$.

e) $(-4)^{-2} = \frac{1}{(-4)^2}$ Use the definition of a negative exponent.

$= \frac{1}{16}$ $(-4)^2 = (-4)(-4) = 16$

f) $5x^{-3} = 5 \cdot \frac{1}{x^3}$ The negative exponent is only on x.

$= \frac{5}{x^3}$

Note that examples 2c) and 2f) could not be simplified if $x = 0$ (0^0 is not defined). For the remainder of this chapter, assume that all variables represent nonzero real numbers.

Writing numbers in scientific notation

Scientific notation is a shorthand method used for writing very large or very small numbers. 8,000,000 can be written as 8×10^6 in scientific notation. The rules for writing numbers in scientific notation follow.

To Write a Number in Scientific Notation:

1. Move the decimal point to the right of the first nonzero digit. Count the number of places you moved the decimal point.
2. Multiply the numbers from step 1 times 10 raised to + or – the number of places you moved the decimal point.
   Use + if you moved the decimal to the left.
   Use – if you moved the decimal to the right.

EXAMPLE 3 Write each number in scientific notation.

a) 836,000

b) 0.000658

SOLUTION 3

a) 836,000 — Given number.

836000. — Move the decimal point to the right of 8. The decimal point moved 5 places.

$= 8.36 \times 10^5$ — Use +5 (or 5) as the exponent on 10 since the decimal moved *to the left* 5 places.

b) 0.000658

0.000658 — Move the decimal to the right of 6. The decimal point moved 4 places.

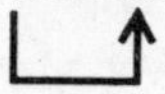

$= 6.58 \times 10^{-4}$ — Use –4 as the exponent on 10 since the decimal moved *to the right* 4 places.

Writing numbers in expanded form

If you own a calculator and have entered very large or very small numbers, perhaps your display has looked like

$$4.26^{12} \text{ or } 4.26 \text{ EE } 12$$

If so, the display is showing a number in scientific notation. To write that number in expanded form (or regular notation), follow these rules.

To Write a Number in Expanded Form:

1. If the exponent on 10 is positive (+), move the decimal point to the right the same number of places as the exponent. Add zeros as necessary.
2. If the exponent on 10 is negative (–), move the decimal point to the left the same number of places as the exponent. Add zeros as necessary.

EXAMPLE 4 Write each number in expanded form.

a) 1.78×10^2

b) 4.26×10^{12}

c) 3.54×10^{-4}

d) 2.19×10^{-1}

SOLUTION 4

a) 1.78×10^2

$= 178$

Move the decimal point 2 places to the right.

b) 4.26×10^{12}

$= 4260000000000$

Move the decimal point 12 places to the right. Add necessary zeros.

c) 3.54×10^{-4}

$= 0.000354$

Move the decimal point 4 places to the left.

d) $2.19 \times 10^{-1} = 0.219$

Move the decimal point 1 place to the left.

3.2 PROPERTIES OF EXPONENTS

The following list of exponent laws can be used for integer exponents, m and n.

$a^m a^n = a^{m+n}$

When multiplying, if bases are the same, add the exponents.

$(a^m)^n = a^{mn}$

When raising a power to a power, multiply the exponents.

$(ab)^m = a^m b^m$

When raising a product to a power, raise each factor to the power.

$\left(\frac{a}{b}\right)^m = \frac{a^m}{b^m}, b \neq 0, b \neq 0$

When raising a fraction to a power, raise numerator and denominator to the power.

$\frac{a^m}{a^n} = a^{m-n}, a \neq 0$

When dividing, if the bases are the same, subtract the exponent in the denominator from the exponent in thenumerator.

Note in the first law that the bases must be the same in order to add the exponents.

EXAMPLE 5 Simplify each of the following expressions. Write all answers with positive exponents.

a) $2^3 \cdot 2^{-6}$

b) $x^{-3} \cdot x^{-4} \cdot x^2$

c) $(3^2)^{-4}$

d) $(y^{-2})^{-3}$

e) $(4x)^2$

f) $(5p^{-3})^2$

SOLUTION 5

a) $2^3 \cdot 2^{-6} = 2^{3+(-6)} = 2^{-3}$ When multiplying and bases are equal, add the exponents.

$= (\frac{1}{2})^3 = \frac{1}{8}$

b) $x^{-3} \cdot x^{-4} \cdot x^2 = x^{-3+(-4)+2}$ When multiplying and bases are equal, add the exponents.

$= x^{-5}$ You need to write the answer with a positive exponent.

$= \frac{1}{x^5}$ Use the definition of negative exponents to rewrite x^{-5}.

c) $(3^2)^{-4} = 3^{2(-4)}$ To raise a power to a power, multiply the exponents.

$= 3^{-8}$ You need to write the answer with a positive exponent.

$= \frac{1}{3^8}$ Use the definition of negative exponents to rewrite 3^{-8}.

d) $(y^{-2})^{-3} = y^{-2(-3)}$ To raise a power to a power, multiply the exponents.

$= y^6$

e) $(4x)^2 = 4^2x^2$ To raise a product to a power, raise each factor to the power.

$= 16x^2$ Simplify $4^2 = 16$.

f) $(5p^{-3})^2 = 5^2p^{-3(2)}$ To raise a product to a power, raise each factor to the power.

$= 25p^{-6}$ Use the definition of negative exponents to rewrite p^{-6}.

$$= \frac{25}{p^6}$$

EXAMPLE 6 Simplify. Write all answers with positive exponents.

a) $\left(\frac{2}{3}\right)^{-2}$

b) $\frac{x^{-4}}{x^3}$

c) $(a^3b^{-2})^{-4}$

d) $\frac{5^{-3}m^4n^{-7}}{5^4m^{-2}n}$

e) $\left(\frac{7p}{q^2}\right)^2\left(\frac{5p^3}{q^{-4}}\right)^{-1}$

SOLUTION 6

a) $\left(\frac{2}{3}\right)^{-2} = \frac{2^{-2}}{3^{-2}}$

To raise a fraction to a power, raise the numerator and denominator to the power.

$$= \frac{\frac{1}{2^2}}{\frac{1}{3^2}}$$

Use the definition of a negative exponent to rewrite

$2^{-2} = \frac{1}{2^2}$ and $3^{-2} = \frac{1}{3^2}$.

$$= \frac{\frac{1}{4}}{\frac{1}{9}}$$

The fraction bar is a division symbol.

$$= \frac{1}{4} \div \frac{1}{9} = \frac{1}{4} \cdot \frac{9}{1}$$

Invert and multiply.

$$= \frac{9}{4}$$

Example 6a) provides a demonstration of the technique used to prove:

$$\left(\frac{a}{b}\right)^{-n} = \left(\frac{b}{a}\right)^{n}$$

A nonzero number raised to the negative n equals the reciprocal of the number raised to the n.

b) $\frac{x^{-4}}{x^{3}} = x^{-4-3} = x^{-7}$

The bases are equal so subtract the exponents.

c) $(a^3b^{-2})^{-4} = a^{3(-4)}b^{-2(-4)}$

To raise a product to a power, raise each factor to the power.

$= a^{-12}b^{8}$

You need to write the answer with positive exponents.

$= \frac{1}{a^{12}} \bullet b^8 = \frac{b^8}{a^{12}}$

Rewrite $a^{-12} = \frac{1}{a^{12}}$.

d) $\frac{5^{-3}m^4n^{-7}}{5^4m^{-2}n} = 5^{-3-4}m^{4-(-2)}n^{-7-1}$

Subtract the exponents on equal bases.

$= 5^{-7}m^6n^{-8}$

You need to write the answer with positive exponents.

$= \frac{1}{5^7} \cdot m^6 \cdot \frac{1}{n^8} = \frac{m^6}{5^7n^8}$

Rewrite $5^{-7} = \frac{1}{5^7}$ and $n^{-8} = \frac{1}{n^8}$.

e) $\left(\frac{7p}{q^2}\right)^2 \left(\frac{5p^3}{q^{-4}}\right)^{-1} = \frac{7^2p^2}{q^4} \cdot \frac{q^4}{5p^3}$

Simplify each fraction.

$= \frac{7^2}{p} = \frac{49}{p}$

Simplify $\frac{q^4}{q^4} = 1$ and $\frac{p^2}{p^3} = p^{2-3} = p^{-1} = \frac{1}{p}$.

Exponent laws and scientific notation

We can use our exponent laws to simplify calculations involving scientific notation.

EXAMPLE 7 Use scientific notation to compute each of the following.

a) $(12{,}000{,}000)^2$

b) $\dfrac{0.00009 \times 6400000}{180000 \times 8000}$

SOLUTION 7

a) $(12{,}000{,}000)^2 = (1.2 \times 10^7)^2$ — Write 12,000,000 in scientific notation.

$= (1.2)^2(10^{7 \cdot 2})$ — To raise a product to a power, raise each factor to the power.

$= 1.44 \times 10^{14}$

b) $\dfrac{0.00009 \times 6400000}{180000 \times 8000}$

$= \dfrac{9 \times 10^{-5} \times 6.4 \times 10^6}{1.8 \times 10^5 \times 8 \times 10^3}$ — Write each number in scientific notation.

$= \dfrac{9 \times 6.4 \times 10^{-5} \times 10^6}{1.8 \times 8 \times 10^5 \times 10^3}$ — Use the commutative property to change the order.

$= \dfrac{57.6 \times 10^1}{14.4 \times 10^8}$ — Add exponents on 10 in the numerator and the denominator.

$= 4 \times 10^{-7}$ — Divide $\dfrac{57.6}{14.4} = 4$. Subtract exponents on equal bases of 10.

3.3 ADDING, SUBTRACTING, AND EVALUATING POLYNOMIALS

Polynomial vocabulary

The following table presents language we will use in our work with polynomials.

VOCABULARY	MEANING	EXAMPLE
Term	A number or a product of a number and one or more variables.	In $6x^2 + 5x + 8$ there are 3 terms: $6x^2$, $5x$, and 8.
Numerical coefficient	Numerical factor in a term.	4 is the numerical coefficient in $4x^3$. -2 is the numerical coefficient in $-2xy^2$.
Polynomial	Expression containing a finite sum of terms.	$4x$ $x^2 + 2x = 1$
Monomial	Polynomial with 1 term.	6 $2x$
Binomial	Polynomial with 2 terms.	$x^2 - 9$ $2x + 6$
Trinomial	Polynomial with 3 terms.	$x^2 + 2x + 1$
Degree of a term	Sum of exponents of the variables within the term.	Degree of $6x^2y^3$ is 5. Degree of $2x = 2x^1$ is 1. Degree of $5 = 5x^0$ is 0.
Degree of a polynomial in 1 variable	Highest degree of any term.	Degree of $4x^3 - 2x$ is 3.
Similar terms or like terms	Terms that have the same variables and exponents.	$4x^3$ and $-6x^3$ $6a$ and $2a$

EXAMPLE 8 Identify each polynomial as a monomial, binomial, or trinomial.

a) $3x - 5$

b) $4x^2 + 6x - 3$

c) $2x^2yz$

SOLUTION 8

a) $3x - 5$ is a binomial.

b) $4x^2 + 6x - 3$ is a trinomial.

c) $2x^2yz$ is a monomial.

EXAMPLE 9 Find the degree of each polynomial.

a) $3x^2 - 2x - 1$

b) $8y^5 - 6y^{12} + 4y$

c) $4x$

d) 7

SOLUTION 9

a) $3x^2 - 2x - 1$ has degree 2. — The highest exponent is 2.

b) $8y^5 - 6y^{12} + 4y$ has degree 12. — The highest exponent is 12.

c) $4x = 4x^1$ has degree 1. — No written exponent means an exponent of 1.

d) $7 = 7x^0$ has degree 0. — The degree of any constant is 0.

Descending order

We often write polynomials in one variable in **descending order**, that is, with the highest degree term first and decreasing degree terms in order from left to right.

EXAMPLE 10 Write each polynomial in descending order.

a) $5 + x^2 - 4x$

b) $2y^3 - 14y^5 + 6y + 8y^6 - 3$

SOLUTION 10

a) $5 + x^2 - 4x = x^2 - 4x + 5$ — Reorder the terms with the highest exponent first.

b) $2y^3 - 14y^5 + 6y + 8y^6 - 3 =$
$8y^6 - 14y^5 + 2y^3 + 6y - 3$

Notice that the exponents are decreasing from left to right.

Adding polynomials

Recall from our work with the distributive property that only similar terms can be added.

To Add Polynomials:
1. Group similar terms together, making use of the commutative and associative properties.
2. Add the numerical coefficients of similar terms.

EXAMPLE 11 Add.
a) $(3x^2 - 4x + 7) + (-5x^2 + x - 11)$
b) $(4y^5 - 6y^3 + 2y - 8) + (8y^5 - 2y^3 - y^2 + 3)$

SOLUTION 11

a) $(3x^2 - 4x + 7) + (-5x^2 + x - 11) =$
$3x^2 - 5x^2 - 4x + x + 7 - 11 =$ Group similar terms together.

$-2x^2 - 3x - 4$

b) $(4y^5 - 6y^3 + 2y - 8) + (8y^5 - 2y^3 - y^2 + 3) =$
$4y^5 + 8y^5 - 6y^3 - 2y^3 - y^2 + 2y - 8 + 3 =$ Group similar terms together.

$12y^5 - 8y^3 - y^2 + 2y - 5$ Add numerical coefficients of similar terms. Note that $-y^2$ and $2y$ did not combine with any other terms.

Adding polynomials vertically

The previous addition examples were done horizontally. Sometimes it's convenient to work problems vertically, lining up the similar terms.

EXAMPLE 12 Add these polynomials vertically.

a) $(3x^2 - 4x + 7) + (-5x^2 + x - 11)$

b) $(4y^5 - 6y^3 + 2y - 8) + (8y^5 - 2y^3 - y^2 + 3)$

SOLUTION 12

a)
$$\begin{array}{r} 3x^2 - 4x + 7 \\ 5x^2 + x - 11 \\ \hline -2x^2 - 3x - 4 \end{array}$$

Line up the polynomials vertically.

Add the numerical coefficients of similar terms.

b)
$$\begin{array}{rrrrr} 4y^5 - 6y^3 & & + 2y - 8 \\ 8y^5 - 2y^3 - y^2 & & + 3 \\ \hline 12y^5 - 8y^3 - y^2 + 2y - 5 \end{array}$$

Line up the polynomials vertically, leaving spaces for missing powers.

Add the numerical coefficients of similar terms.

Subtracting polynomials

To subtract two polynomials, change all the signs of the terms of the polynomial to be subtracted, and then add.

EXAMPLE 13 Subtract.

a) $(8x^2 - 6x + 4) - (3x^2 - 7x - 5)$

b) $(6p^4 - 5p^3 - 2p) - (-7p^4 - 6p + 2)$

SOLUTION 13

a) $(8x^2 - 6x + 4) - (3x^2 - 7x - 5) =$

$(8x^2 - 6x + 4) + (-3x^2 + 7x + 5) =$

Change all the signs in the

second polynomial.

$(8x^2 - 3x^2) + (-6x + 7x) + (4 + 5) =$ Group similar terms.

$5x^2 + x + 9$ Combine similar terms.

b) $(6p^4 - 5p^3 - 2p) - (-7p^4 - 6p + 2) =$

$(6p^4 - 5p^3 - 2p) + (7p^4 + 6p - 2) =$ Change all the signs in the second polynomial.

$(6p^4 + 7p^4) + (-5p^3) + (-2p + 6p) + (-2) =$ Group similar terms.

$13p^4 - 5p^3 + 4p - 2$ Combine similar terms.

We can set up subtraction problems vertically, lining up similar terms.

EXAMPLE 14 Subtract these polynomials vertically.

a) $(8x^2 - 6x + 4) - (3x^2 - 7x - 5)$

b) $(6p^4 - 5p^3 - 2p) - (-7p^4 - 6p + 2)$

SOLUTION 14

a)
$$\begin{array}{r} 8x^2 - 6x + 4 \\ \underline{-(3x^2 - 7x - 5)} \end{array}$$
Line up the polynomials vertically.

$$\begin{array}{r} 8x^2 - 6x + 4 \\ \underline{-3x^2 + 7x + 5} \\ 5x^2 + x + 9 \end{array}$$
Change all the signs in the second polynomial and add.

b)
$$\begin{array}{l} 6p^4 - 5p^3 - 2p \\ \underline{-(-7p^4 \qquad\quad - 6p + 2)} \end{array}$$
Line up the polynomials vertically.

$$\begin{array}{l} 6p^4 - 5p^3 - 2p \\ \underline{+7p^4 \qquad\quad + 6p - 2} \\ 13p^4 - 5p^3 + 4p - 2 \end{array}$$
Change all the signs in the second polynomial and add.

Evaluating polynomials

Polynomials are often labeled with capital letters, such as $P(x) = 4x^2 - 8x + 4$ (read P of x). Evaluating a polynomial means finding the value of the polynomial for a given value of the variable.

EXAMPLE 15 Let $P(x) = 4x^2 - 8x + 4$. Find each of the following.

a) $P(3)$

b) $P(-2)$

SOLUTION 15

a) $P(3) = 4(3)^2 - 8(3) + 4$	Substitute 3 for x in $P(x)$.
$= 4(9) - 8(3) + 4$	Simplify exponents.
$= 36 - 24 + 4$	Multiply.
$= 16$	Add and subtract in order from left to right.

b) $P(-2) = 4(-2)^2 - 8(-2) + 4$	Substitute -2 for x in $P(x)$.
$= 4(4) - 8(-2) + 4$	Simplify exponents.
$= 16 + 16 + 4$	Multiply.
$= 36$	Add.

3.4 MULTIPLYING POLYNOMIALS

Multiplying by a monomial

To multiply by a monomial, multiply the numerical coefficients, and add exponents on equal bases. Use the distributive property to multiply a monomial times a polynomial with more than one term.

EXAMPLE 16 Multiply.

a) $(-4x^2)(3x^4)$

b) $3x^2(5x^3 - 4x^2 + 2x - 1)$

c) $2a^2b^3(-4a^3 + 5a^2b - 6ab^2 + 2b^3)$

SOLUTION 16

a) $(-4x^2)(3x^4) =$

$-4 \cdot 3x^{2+4} =$ Multiply coefficients and add exponents on equal bases.

$-12x^6$

b) $3x^2(5x^3 - 4x^2 + 2x - 1) =$

$3x^2(5x^3) + 3x^2(-4x^2) + 3x^2(2x) + 3x^2(-1) =$ Use the distributive property to multiply each term of the second polynomial by $3x^2$.

$15x^5 - 12x^4 + 6x^3 - 3x^2$ Multiply the coefficients, add the exponents on equal bases.

c) $2a^2b^3(-4a^3 + 5a^2b - 6ab^2 + 2b^3) =$

$2a^2b^3(-4a^3) + 2a^2b^3(5a^2b) + 2a^2b^3(-6ab^2) + 2a^2b^3(2b^3) =$

Use the distributive property to multiply each term of the second polynomial by $2a^2b^3$.

$-8a^5b^3 + 10a^4b^4 - 12a^3b^5 + 4a^2b^6$ Multiply the coefficients, add the exponents on equal bases.

Multiplying two polynomials

Using the distributive property leads to the following rules for multiplying two polynomials.

To Multiply Two Polynomials:
1. Multiply each term of the second polynomial by each term of the first polynomial.
2. Add any similar terms.

EXAMPLE 17 Multiply.

a) $(2a + 4)(3a - 5)$

b) $(3m - 2)(4m^2 - 6m - 3)$

SOLUTION 17

a) $(2a + 4)(3a - 5) =$

$2a(3a) + 2a(-5) + 4(3a) + 4(-5) =$ Multiply each term of the second polynomial by each term of the first polynomial.

$6a^2 - 10a + 12a - 20 =$ Multiply coefficients, add the exponents on equal bases.

$6a^2 + 2a - 20$ Add similar terms: $-10a + 12a = 2a$.

b) $(3m - 2)(4m^2 - 6m - 3) =$

$3m(4m^2) + 3m(-6m) + 3m(-3) + (-2)(4m^2) + (-2)(-6m) + (-2)(-3) =$

Multiply each term of the second polynomial by each term of the first polynomial.

$12m^3 - 18m^2 - 9m - 8m^2 + 12m + 6 =$

Multiply coefficients, add the exponents on equal bases.

$12m^3 + (-18m^2 - 8m^2) + (-9m + 12m) + 6 =$ Group similar terms together.

$12m^3 - 26m^2 + 3m + 6$ Combine similar terms.

Multiplying polynomials vertically

Just as we could line up polynomials vertically to add and subtract, we can also multiply polynomials vertically. This technique is especially useful for multiplications involving polynomials with three or more terms.

EXAMPLE 18 Find the product of $x^2 - 2xy + y^2$ and $x^2 + 2xy + y^2$.

SOLUTION 18

$x^2 - 2xy + y^2$	Line up the polynomials
$x^2 + 2xy + y^2$	vertically.
$x^2y^2 - 2xy^3 + y^4$ ←	$y^2(x^2 - 2xy + y^2)$
$2x^3y - 4x^2y^2 + 2xy^3$ ←	$2xy(x^2 - 2xy + y^2)$
$x^4 - 2x^3y + x^2y^2$ ←	$x^2(x^2 - 2xy + y^2)$
$x^4 - 2x^2y^2 + y^4$	Combine similar terms.

Multiplying using FOIL

Several techniques can help speed up multiplication. The first technique is used for multiplying two binomials.

> **To Multiply Two Binomials using FOIL:**
> 1. Multiply the **F**irst two terms.
> 2. Multiply the **O**uter two terms.
> 3. Multiply the **I**nner two terms.
> 4. Multiply the **L**ast two terms.
> 5. Add the four terms, combining similar terms wherever possible.

EXAMPLE 19 Multiply using FOIL.

a) $(2x - 5)(3x + 7)$

b) $(4a - 3)(2a - 1)$

c) $(3x + 5y)(4x - 3y)$

SOLUTION 19

a) $(2x - 5)(3x + 7) =$

F: $2x(3x) = 6x^2$	Multiply the first two terms.
O: $2x(7) = 14x$	Multiply the outer two terms.
I: $-5(3x) = -15x$	Multiply the inner two terms.
L: $-5(7) = -35$	Multiply the last two terms.
$= 6x^2 + 14x - 15x - 35$	Add the four terms.
$= 6x^2 - x - 35$	Combine similar terms.

b) $(4a - 3)(2a - 1) =$

F: $4a(2a) = 8a^2$	Multiply the first two terms.
O: $4a(-1) = -4a$	Multiply the outer two terms.
I: $-3(2a) = -6a$	Multiply the inner two terms.
L: $-3(-1) = 3$	Multiply the last two terms.
$= 8a^2 - 4a - 6a + 3$	Add the four terms.
$= 8a^2 - 10a + 3$	Combine similar terms.

c) $(3x + 5y)(4x - 3y)$

F: $3x(4x) = 12x^2$	Multiply the first two terms.
O: $3x(-3y) = -9xy$	Multiply the outer two terms.
I: $5y(4x) = 20xy$	Multiply the inner two terms.
L: $5y(-3y) = -15y^2$	Multiply the last two terms.
$= 12x^2 - 9xy + 20xy - 15y^2$	Add the four terms.
$= 12x^2 + 11xy - 15y^2$	Combine similar terms.

Squaring a binomial

Recall that squaring a quantity means multiplying it times itself. Although we can square a binomial by writing it twice and using FOIL, the following rules will help you shorten the process.

> **To Square a Binomial:**
> 1. Square the first term.
> 2. Multiply 2 times the first term times the last term.
> 3. Square the last term.
> 4. Add these terms together.

EXAMPLE 20 Multiply.

a) $(a + 5)^2$

b) $(m - 3)^2$

c) $(2x + 3y)^2$

d) $(4r - 5s)^2$

SOLUTION 20

a) $(a + 5)^2 =$

$(a)^2 = a^2$	Square the first term.
$2(a)(5) = 10a$	Multiply 2 times the first term (a) times the last term (5).
$(5)^2 = 25$	Square the last term.
$= a^2 + 10a + 25$	Add the terms together.

b) $(m - 3)^2 =$

$(m)^2 = m^2$	Square the first term.
$2(m)(-3) = -6m$	Multiply 2 times the first term (m) times the last term (–3).
$(-3)^2 = 9$	Square the last term.
$= m^2 - 6m + 9$	Add the terms together.

c) $(2x + 3y)^2 =$

$(2x)^2 = 4x^2$	Square the first term.
$2(2x)(3y) = 12xy$	Multiply 2 times the first term ($2x$) times the last term ($3y$).
$(3y)^2 = 9y^2$	Square the last term.

$= 4x^2 + 12xy + 9y^2$ — Add the terms together.

d) $(4r - 5s)^2$

$(4r)^2 = 16r^2$ — Square the first term.

$2(4r)(-5s) = -40rs$ — Multiply 2 times the first term $(4r)$ times the last term $(-5s)$.

$(-5s)^2 = 25s^2$ — Square the last term.

$= 16r^2 - 40rs + 25s^2$ — Add the terms together.

Multiplying the sum and difference of two like terms

When the sum and difference of two like terms are multiplied using FOIL, the inner and outer terms sum to 0. This fact leads to the rules for multiplying the sum and difference of two like terms.

> **To Multiply the Sum and Difference of Two Like Terms:**
> 1. Square the first term.
> 2. Square the last term.
> 3. Subtract these two terms.

EXAMPLE 21 Multiply.

a) $(x - y)(x + y)$

b) $(3a - 4b)(3a + 4b)$

c) $[(x + y) + 5][(x + y) - 5]$

SOLUTION 21

a) $(x - y)(x + y) =$

$(x)^2 = x^2$ — Square the first term.

$(y)^2 = y^2$ — Square the last term.

$= x^2 - y^2$ — Subtract.

b) $(3a - 4b)(3a + 4b)$

$(3a)^2 = 9a^2$ — Square the first term.

$(4b)^2 = 16b^2$ — Square the last term.

$= 9a^2 - 16b^2$ — Subtract.

c) $[(x+y)+5][(x+y)-5]$

$(x+y)^2 = x^2 + 2xy + y^2$ — Use the rules for squaring a binomial.

$(5)^2 = 25$ — Square the last term.

$= x^2 + 2xy + y^2 - 25$ — Subtract.

3.5 GREATEST COMMON FACTOR AND FACTORING BY GROUPING

The remaining sections in this chapter are about factoring, that is, breaking numbers and polynomials into factors. We'll begin by factoring integers.

Prime factoring

A **prime number** is an integer greater than one that can only be evenly divided by itself and 1. For example,

$$2, 3, 5, 7, 11, 13, 17, 19, 23, \ldots$$

are prime numbers. We can **prime factor** an integer by writing it as a product of prime numbers, for example $6 = 2 \cdot 3$. To prime factor larger numbers, begin dividing by the smallest possible prime and continue until the quotient is a prime number.

EXAMPLE 22 Prime factor each number.

a) 72

b) 48

c) 36

SOLUTION 22

a) $72 = 2 \cdot 36$ — 2 is the smallest prime that divides into 72.

$= 2 \cdot 2 \cdot 18$	Divide by 2 again.
$= 2 \cdot 2 \cdot 2 \cdot 9$	Divide by 2 again.
$= 2 \cdot 2 \cdot 2 \cdot 3 \cdot 3$	3 is the smallest prime number that divides into 9.
$= 2^3 \cdot 3^2$	Write your answer using exponents.
b) $48 = 2 \cdot 24$	2 is the smallest prime that divides into 48.
$= 2 \cdot 2 \cdot 12$	Divide by 2.
$= 2 \cdot 2 \cdot 2 \cdot 6$	Divide by 2.
$= 2 \cdot 2 \cdot 2 \cdot 2 \cdot 3$	Divide by 2.
$= 2^4 \cdot 3$	Write your answer using exponents.
c) $36 = 2 \cdot 18$	2 is the smallest prime that divides into 36.
$= 2 \cdot 2 \cdot 9$	Divide by 2.
$= 2 \cdot 2 \cdot 3 \cdot 3$	3 is the smallest prime number that divides into 9.
$= 2^2 \cdot 3^2$	Write your answer using exponents.

Finding the greatest common factor

The greatest common factor (GCF) of a list of terms is the largest factor (including numerical coefficient and variables) that evenly divides into each term.

> **Finding the GCF:**
> 1. Prime factor each numerical coefficient and write the factors using exponents.
> 2. The GCF is the product of bases that are common to all terms, raised to the lowest exponent that appears in any of the terms.

EXAMPLE 23 Find the GCF.

a) 72, 48 and 36

b) $4x^3, 6x^5y^2$

SOLUTION 23

a) $72 = 2^3 \cdot 3^2$ — Prime factor each term.

$48 = 2^4 \cdot 3$

$36 = 2^2 \cdot 3^2$

$\text{GCF} = 2^2 \cdot 3^1$ — Use bases 2 and 3 because they appear in *all* terms. The lowest exponent on 2 is 2, the lowest exponent on 3 is 1.

b) $4x^3 = 2 \cdot x^3$ — Prime factor the numerical coefficients.

$6x^5y^2 = 2 \cdot 3 \cdot x^5 \cdot y^2$

$\text{GCF} = 2^1x^3$ — Use bases 2 and x because they appear in both terms. The lowest exponent on 2 is 1, the lowest exponent on x is 3.

Factoring out the GCF

In the beginning of this chapter we multiplied polynomials. We will now reverse the multiplication process, called factoring. The first step in any factoring problem is to factor out the greatest common factor.

To Factor Out the GCF:

1. Write the GCF of all the terms followed by a set of parentheses. Leave space inside the parentheses for the same number of terms found in the original polynomial.
2. Divide the GCF into *each* term of the original polynomial and write each quotient inside the parentheses.

EXAMPLE 24 Factor out the GCF.

a) $4x^3 + 6x^5y^2$

b) $6x^2 + 5$

c) $72r^6s^2 + 48r^4s^2 - 36r^3s^2$

d) $8(x + y) - 5a(x + y)$

e) $-6x^3 + 8x - 4$

f) $24x^2y^2 - 24$

SOLUTION 24

a) $4x^3 + 6x^5y^2$

$2x^3(\qquad\qquad)$ — Write the GCF in front of a set of parentheses. Leave space in the parentheses for 2 terms.

$= 2x^3(2 + 3x^2y^2)$ — Divide the GCF into each term. Write the quotients inside the parentheses.

b) $6x^2 + 5$ — The GCF of $6x^2$ and 5 is 1. When the GCF is 1, we leave the original polynomial in its given form.

c) $72r^6s^2 + 48r^4s^2 - 36r^3s^2$

$12r^3s^2(\qquad\qquad)$ — Write the GCF followed by a set of parentheses with space for 3 terms.

$= 12r^3s^2(6r^3 + 4r - 3)$ — Divide the GCF into each term. Write the quotients inside the parentheses.

d) $8(x + y) - 5a(x + y)$ — Here the GCF is $(x + y)$.

$(x + y)(\qquad\qquad)$ — Write the GCF followed by a set of parentheses with space for 2 terms.

$= (x + y)(8 - 5a)$ — Divide the GCF into each term. Write the quotients inside the parentheses.

e) $-6x^3 + 8x - 4$

Method 1.

When the leading coefficient is negative,some instructors prefer to factor out that negative.

$-2(\qquad\qquad)$ — Write –2 followed by a set of parentheses with space for 3 terms.

$= -2(3x^2 - 4x + 2)$ — Note that each sign changes within the parentheses.

Method 2.

Sometimes it will be necessary to factor out the positive GCF.

$2(\qquad\qquad)$ — Write 2 followed by a set of parentheses with space for 3 terms.

$= 2(-3x^3 + 4x - 2)$

f) $24x^2y^2 - 24$

$24(\qquad\qquad)$ — Write the GCF followed by a set of parentheses with space for 2 terms.

$= 24(x^2y^2 - 1)$ — Divide the GCF into each term. Be careful here to write the 1.

Note that when finding common factors there will always be the same number of terms in the parentheses as there were in the original polynomial. Also note that factoring problems can be checked by multiplying the factors. However, you must be careful to factor out the *greatest* common factor.

$4x + 8y = 4(x + 2y)$	Factored completely.
$4x + 8y = 2(2x + 4y)$	Not factored completely; GCF is 4.

Though both answers check, only the first is considered to be factored completely.

Factoring by grouping

In this section we will concentrate on factoring four term polynomials that can be grouped as two terms plus two terms.

To Factor by Grouping:

1. Group together the first two terms and factor out the GCF. Group together the last two terms and factor out the GCF.
2. If the expressions in the parentheses are equal, factor out that expression as the GCF.
3. If the expressions in the parentheses are opposites, factor out –1 from one set of parentheses and go to step 2. If the expressions in the parentheses are not equal (and not opposites), change the order of the terms and go back to step 1.

EXAMPLE 25 Factor.

a) $ax + xy + 2a + 2y$

b) $x^2y^2 - 2y^2 + 6x^2 - 12$

c) $8x + 8y - 5ax - 5ay$

d) $r^2s^2 + 3t^2 + r^2t^2 + 3s^2$

SOLUTION 25

a) $ax + xy + 2a + 2y$

$= (xa + xy) + (2a + 2y)$ — Group together the first two terms and the last two terms.

$= x(a + y) + 2(a + y)$ — Factor out the GCF from each pair of parentheses.

$= (a + y)(x + 2)$ — $(a + y)$ is now the GCF and can be factored out.

b) $x^2y^2 - 2y^2 + 6x^2 - 12$

$= (x^2y^2 - 2y^2) + (6x^2 - 12)$ — Group together the first two terms and the last two terms.

$= y^2(x^2 - 2) + 6(x^2 - 2)$ — Factor out the GCF from each pair of parentheses.

$= (x^2 - 2)(y^2 + 6)$ — $(x^2 - 2)$ is now the GCF and can be factored out.

c) $8x + 8y - 5ax - 5ay$

$= (8x + 8y) + (-5ax - 5ay)$ Group together the first two terms and the last two terms.

$= 8(x + y) + 5a(-x - y)$

The terms in the parentheses are opposites.

Factor out -1 from the second parentheses:

$= 8(x + y) - 5a(x + y)$ Expressions in the parentheses are equal.

$= (x + y)(8 - 5a)$ Factor out the GCF $(x + y)$.

d) $r^2s^2 + 3t^2 + r^2t^2 + 3s^2$

$= (r^2s^2 + 3t^2) + (r^2t^2 + 3s^2)$ Group together the first two terms and the last two terms.

There are no common factors in either set of parentheses. Change the order of the terms and try again.

$= r^2s^2 + r^2t^2 + 3t^2 + 3s^2$ Change the order.

$= (r^2s^2 + r^2t^2) + (3t^2 + 3s^2)$ Group together the first two terms and the last two terms.

$= r^2(s^2 + t^2) + 3(t^2 + s^2)$ Note that $s^2 + t^2 = t^2 + s^2$, so you can common factor.

$= (s^2 + t^2)(r^2 + 3)$ This can be written also as $(t^2 + s^2)(r^2 + 3)$.

3.6 FACTORING TRINOMIALS

Previously we used FOIL to multiply two binomials such as

$(x + 1)(x + 4) = x^2 + 4x + 1x + 4 = x^2 + 5x + 4$

$(2x - 3)(x + 5) = 2x^2 + 10x - 3x - 15 = 2x^2 + 7x - 15$

In each case, the product was a trinomial. In this section, we wish to factor trinomials, that is, to start with $x^2 + 5x + 4$ and to end with $(x + 1)(x + 4)$.

Factoring trinomials when a = 1

In the trinomial $ax^2 + bx + c$, a is called the leading coefficient. When $a = 1$, or when $a = 1$ after common factoring, we can factor $ax^2 + bx + c$ using the following rules:

To Factor $x^2 + bx + c$:

1. List all pairs of factors of c.
2. Find a pair of factors whose sum equals b. If there is no such pair, the trinomial does not factor (called **prime**).
3. Write two sets of parentheses $(x \_\_\_\_)(x \_\_\_\_)$.
4. Fill in the blanks with the pair from step 2.
5. Check your answer using FOIL.

EXAMPLE 26 Factor each trinomial.

a) $x^2 + 5x + 4$

b) $m^2 - 2m - 63$

c) $a^2 - 7ab + 12b^2$

d) $p^2 - 2p + 15$

e) $2y^2 + 4y - 30$

SOLUTION 26

a) $x^2 + 5x + 4$

$c = +4$		List all pairs of factors of +4.
+1	+4	Find a pair whose sum equals 5.
+2	+2	
−1	−4	
−2	−2	

$(x \_\_\_\_)(x \_\_\_\_)$ Set up a pair of parentheses.

$(x + 1)(x + 4)$ Fill in the blanks.

Check:

$(x+1)(x+4) =$ $x^2+4x+1x+4 =$ x^2+5x+4	Check your answer using FOIL.

b) $m^2 - 2m - 63$

$c = -63$	List all pairs of factors of –63.
$+1 \quad -63$ $+3 \quad -21$ $\mathbf{+7 \quad -9}$ $-1 \quad +63$ $-3 \quad +21$ $-7 \quad +9$	Find a pair whose sum equals –2.
$(m \_\_\_\_)(m \_\_\_\_)$	Set up a pair of parentheses.
$(m+7)(m-9)$	Fill in the blanks.

Check:

$(m+7)(m-9) =$ $m^2 - 9m + 7m - 63 =$ $m^2 - 2m - 63$	Check your answer using FOIL.

c) $a^2 - 7ab + 12b^2$

$c = +12b^2$	List all pairs of factors of +12.
$+1b \quad +12b$ $+2b \quad +6b$ $+3b \quad +4b$ $-1b \quad -12b$ $-2b \quad -6b$ $\mathbf{-3b \quad -4b}$	Find a pair whose sum equals –7.
$(a \_\_\_\_)(a \_\_\_\_)$	Set up a pair of parentheses.
$(a-3b)(a-4b)$	Fill in the blanks.

Check:

$(a-3b)(a-4b) =$ $a^2 - 4ab - 3ab + 12b^2 =$ $a^2 - 7ab + 12b^2$	Check your answer using FOIL.

d) $p^2 - 2p + 15$

$c = +15$		List all pairs of factors of +15.
+1	+15	
+3	+ 5	
–1	–15	
–3	– 5	

There is no pair of factors whose sum equals –2. This trinomial is prime.

e) $2y^2 + 4y - 30 = 2(y^2 + 2y - 15)$ — Factor out the GCF.

Now factor $y^2 + 2y - 15$.

$c = -15$		List all pairs of factors of –15.
+1	– 15	Find a pair whose sum equals +2.
+3	– 5	
–1	+15	
–3	**+ 5**	

$= 2(y - 3)(y + 5)$ — Be sure to write the GCF in your answer.

Check:

$2(y - 3)(y + 5) =$

$2(y^2 + 5y - 3y - 15) =$ — Use FOIL to multiply the binomials.

$2(y^2 + 2y - 15)$ — Combine similar terms.

$2y^2 + 4y - 30$ — Use the distributive property.

Factoring trinomials when a ≠ 1

When the leading coefficient, a, does not equal 1, there are two techniques used for factoring $ax^2 + bx + c$. The factoring by grouping technique uses a four term polynomial equal to the original trinomial and then factoring by grouping to factor $ax^2 + bx + c$. The trial and error process uses possible pairs of binomial factors to factor $ax^2 + bx + c$. Both techniques require practice, but either technique can be used to factor $ax^2 + bx + c$.

Factoring $ax^2 + bx + c$ by grouping

The following steps assume you have already factored out the GCF.

To Factor $ax^2 + bx + c$ by Grouping:
1. Multiply $a \cdot c$.
2. Find a pair of factors of $a \cdot c$ whose sum equals b.
3. Rewrite $ax^2 + bx + c$ by replacing bx with the pair found in step 3.
4. Factor this new four term polynomial by grouping.

EXAMPLE 27 Factor.

a) $12x^2 - 11x - 5$

b) $3t^2 + 8t - 35$

c) $3a^2 - 10ab + 8b^2$

SOLUTION 27

a) $12x^2 - 11x - 5$

$12x^2 \cdot -5 = -60x^2$ Multiply $a \cdot c$.

$-60x^2 = (-15x)(+4x)$ Find a pair of factors of and $-15x + 4x = -11x$ $-60x^2$ whose sum equals $-11x$.

$= 12x^2 - 15x + 4x - 5$ Replace $-11x$ with $-15x + 4x$.

$= 3x(4x - 5) + 1(4x - 5)$ Factor by grouping.

$= (4x - 5)(3x + 1)$

b) $3t^2 + 8t - 35$

$3t^2 \cdot -35 = -105t^2$ Multiply $a \cdot c$.

$-105t^2 = (+15t)(-7t)$ Find a pair of factors of $-150t^2$

$+15t - 7t = +8t$ whose sum equals $+8t$.

$= 3t^2 + 15t - 7t - 35$ Replace $+8t$ with $+15t - 7t$.

$= 3t(t + 5) - 7(t + 5)$ Factor by grouping.

$= (t + 5)(3t - 7)$

c) $3a^2 - 10ab + 8b^2$

$3a^2 \cdot 8b^2 = 24a^2b^2$ — Multiply $a \cdot c$.

$24a^2b^2 = (-4ab)(-6ab)$ — Find a pair of factors of $24a^2b^2$

and $-4ab - 6ab = -10ab$ — whose sum equals $-10ab$.

$= 3a^2 - 4ab - 6ab + 8b^2$ — Replace $-10ab$ with $-4ab - 6ab$.

$= a(3a - 4b) - 2b(3a - 4b)$ — Factor by grouping.

$= (3a - 4b)(a - 2b)$

Factoring by trial and error

To factor $ax^2 + bx + c$ by trial and error we will use the fact that the factors will be a pair of binomials in which the product of the first two terms (F in the FOIL process) must equal ax^2, and that the product of the last two terms (L in the FOIL process) must equal c.

To Factor $ax^2 + bx + c$ by Trial and Error:

1. List positive pairs of factors of ax^2.
2. List pairs of factors of c.
3. List pairs of binomial factors whose first terms are factors of ax^2 and whose last terms are factors of c.
4. Eliminate combinations that have a GCF. If the original trinomial has no GCF, there cannot be a GCF in any of the factors.
5. Use FOIL to multiply the binomials. Stop when you find a pair of factors whose product equals the original trinomial.

Note: You could use positive and negative factors of ax^2, but students generally find it easier to use positive and negative factors of c. If $a < 0$, start by factoring out -1. We'll rework Example 27 using trial and error.

EXAMPLE 28 Factor.

a) $12x^2 - 11x - 5$

b) $3t^2 + 8t - 35$

c) $3a^2 - 10ab + 8b^2$

SOLUTION 28

a) $12x^2 - 11x - 5$

$ax^2 = 12x^2$		c	$= -5$	
$1x$	$12x$	$+1$	-5	List pairs of factors of ax^2 and c.
$2x$	$6x$	-1	$+5$	
$3x$	$4x$			
$4x$	$3x$			
$6x$	$2x$			
$12x$	$1x$			

$(x + 1)(12x - 5)$
$(x - 1)(12x + 5)$
$(2x + 1)(6x - 5)$
$(2x - 1)(6x + 5)$
$(3x + 1)(4x - 5)$
$(3x - 1)(4x + 5)$
$(4x + 1)(3x - 5)$
$(4x - 1)(3x + 5)$
$(6x + 1)(2x - 5)$
$(6x - 1)(2x + 5)$
$(12x + 1)(x - 5)$
$(12x - 1)(x + 5)$

List pairs of binomial factors whose first terms are factors of $12x^2$ and whose last terms are factors of -5.

Using FOIL to multiply these possibilities, we find that

$(3x + 1)(4x - 5) = 12x^2 - 15x + 4x - 5 = 12x^2 - 11x - 5$

Thus

$12x^2 - 11x - 5 = (3x + 1)(4x - 5)$

b) $3t^2 + 8t - 35$

$ax^2 = 3t^2$		c	$= -35$	
$1t$	$3t$	$+1$	-35	List pairs of factors of ax^2 and c.
$3t$	$1t$	-1	$+35$	
		$+5$	-7	
		-5	$+7$	

$(t + 1)(3t - 35)$
$(t - 1)(3t + 35)$

List pairs of binomial factors whose first terms

$(t+5)(3t-7)$
$(t-5)(3t+7)$
$(3t+1)(t-35)$
$(3t-1)(t+35)$
$(3t+5)(t-7)$
$(3t-5)(t+7)$

are factors of $12x^2$ and whose last terms are factors of –5.

Using FOIL to multiply these possibilities, we find that

$(t+5)(3t-7) = 3t^2 - 7t + 15t - 35 = 3t^2 + 8t - 35$

Thus

$3t^2 + 8t - 35 = (t+5)(3t-7)$

c) $3a^2 - 10ab + 8b^2$

$ax^2 = 3a^2$		$c = +8b^2$	
a	$3a$	$+b$	$+8b$
$3a$	a	$-b$	$-8b$
		$+2b$	$+4b$
		$-2b$	$-4b$

List pairs of factors of ax^2 and c.

$(a+b)(3a+8b)$
$(a-b)(3a-8b)$
$(a+2b)(3a+4b)$
$(a-2b)(3a-4b)$
$(3a+b)(a+8b)$
$(3a-b)(a-8b)$
$(3a+2b)(a+4b)$
$(3a-2b)(a-4b)$

List pairs of binomial factors whose first terms are factors of $3a^2$ and whose last terms are factors of $+8b^2$.

Using FOIL to multiply these possibilities, we find that

$(a-2b)(3a-4b) = 3a^2 - 4ab - 6ab + 8b^2 = 3a^2 - 10ab + 8b^2$

Thus

$3a^2 - 10ab + 8b^2 = (a-2b)(3a-4b)$

Choosing methods for factoring

You may find that trial and error works best for problems in which a and/ or c are prime and that grouping works well otherwise. If you have used one method in the past and prefer it, use it. Remember to start every factoring exercise by factoring out the GCF.

EXAMPLE 29 Factor.

a) $6a^3 - 39a^2 + 60a$

b) $14m^4 - 16m^3n - 24m^2n^2$

SOLUTION 29

a) $6a^3 - 39a^2 + 60a = 3a(2a^2 - 13a + 20)$ The GCF is $3a$.

$2a^2 - 13a + 20$

$ax^2 = 2a^2$		$c = +20$	
a	$2a$	$+1$	$+20$
$2a$	a	-1	-20
		$+2$	$+10$
		-2	-10
		$+4$	$+5$
		-4	-5

Since a is prime, let's use trial and error.

$(a + 1)(2a + 20)$ Eliminate since $2a + 20$ has a GCF of 2.

$(a - 1)(2a - 20)$ Eliminate since $2a + 20$ has a GCF of 2.

$(a + 2)(2a + 10)$ Eliminate since $2a + 10$ has a GCF of 2.

$(a - 2)(2a - 10)$ Eliminate since $2a + 10$ has a GCF of 2.

$(a + 4)(2a + 5)$

$(a - 4)(2a - 5)$

$(2a + 1)(a + 20)$

$(2a - 1)(a - 20)$

$(2a + 2)(a + 10)$ Eliminate since $2a + 2$ has a GCF of 2.

$(2a - 2)(a - 10)$ Eliminate since $2a + 2$ has a GCF of 2.

$(2a + 4)(a + 5)$ Eliminate since $2a + 4$ has a GCF of 2.

$(2a - 4)(a - 5)$ Eliminate since $2a + 4$ has a GCF of 2.

Using FOIL to multiply these possibilities we find that

$(a - 4)(2a - 5) = 2a^2 - 5a - 8a + 20 = 2a^2 - 13a + 20$

Thus

$6a^3 - 39a^2 + 60a = 3a(a-4)(2a-5)$ — Be sure to write the GCF in your answer.

b) $14m^4 - 16m^3n - 24m^2n^2 =$

$2m^2(7m^2 - 8mn - 12n^2)$ — The GCF is $2m^2$.

$7m^2 \cdot -12n^2 = -84^2n^2$ — Multiply $a \cdot c$. List pairs of factors of $a \cdot c$.

$1mn$	$-84mn$
$2mn$	$-42mn$
$3mn$	$-28mn$
$4mn$	$-21mn$
$6mn$	$-14mn$

Find a pair whose sum equals $-8mn$.

$7m^2 + 6mn - 14mn - 12n^2$ — Replace $-8mn$ with $+6mn - 14mn$.

$m(7m + 6n) - 2n(7m + 6n)$ — Factor by grouping.

$(7m + 6n)(m - 2n)$

Thus $14m^4 - 16m^3n - 24m^2n^2 = 2m^2(7m + 6n)(m - 2n)$

Be sure to write the GCF in your answer.

Factoring by substitution

Some more complicated polynomials can be factored by the techniques we've discussed by substituting an appropriate variable. Study the following examples.

EXAMPLE 30 Factor.

a) $x^4 - 3x^2 - 18$

b) $5(p + 1)^2 - 17(p + 1) + 6$

SOLUTION 30

a) $x^4 - 3x^2 - 18$

Let $y = x^2$

so $x^4 - 3x^2 - 18 = y^2 - 3y - 18$ — Make the substitution.

$y^2 - 3y - 18 = (y + 3)(y - 6)$ Factor the new trinomial.

But $y = x^2$

so $(y + 3)(y - 6) = (x^2 + 3)(x^2 - 6)$ Substitute x^2 for y.

Thus

$x^4 - 3x^3 - 18 = (x^2 + 3)(x^2 - 6)$

b) $5(p + 1)^2 - 17(p + 1) + 6$

Let $y = p + 1$

so $5(p + 1)^2 - 17(p + 1) + 6 = 5y^2 - 17y + 6$ Make the substitution.

$5y^2 - 17y + 6 = (5y - 2)(y - 3)$ Factor the new trinomial.

But $y = p + 1$

so $(5y - 2)(y - 3) = [5(p + 1) - 2][(p + 1) - 3]$ Substitute x^2 for y.

$= [5p + 5 - 2][p + 1 - 3]$ Simplify inside the brackets.

$= (5p + 3)(p - 2)$

3.7 FACTORING SPECIAL POLYNOMIALS

In this section we will factor several special polynomials. The following table summarizes the formulas we will be using:

Difference of Two Squares	$x^2 - y^2 = (x - y)(x + y)$
Sum of Two Squares	$x^2 + y^2$ does not factor.
Perfect Square Trinomial	$x^2 + 2xy + y^2 = (x + y)^2 = (x + y)(x + y)$ $x^2 - 2xy + y^2 = (x - y)^2 = (x - y)(x - y)$
Sum of Two Cubes	$x^3 + y^3 = (x + y)(x^2 - xy + y^2)$
Difference of Two Cubes	$x^3 - y^3 = (x - y)(x^2 + xy + y^2)$

A difference of two squares factors by the rules stated below. Note that other than a possible GCF a sum of two squares does *not* factor using real numbers.

To Factor a Difference of Two Squares:
1. Write each term as a perfect square $(x)^2 - (y)^2$.
2. Write two sets of parentheses $(\quad + \quad)(\quad - \quad)$.
3. Use each term in parentheses from step 1 to fill in the blanks $(x + y)(x - y)$.
4. Check using FOIL.

EXAMPLE 31 Factor each difference of two squares.

a) $4x^2 - 9y^2$

b) $81p^2 - 49q^2$

c) $(a - 3b)^2 - 16$

SOLUTION 31

a) $4x^2 - 9y^2 = (2x)^2 - (3y)^2$ — $(2x)^2 = 4x^2$. $(3y)^2 = 9y^2$.

$= (\quad + \quad)(\quad - \quad)$ — Write two sets of parentheses.

$= (2x + 3y)(2x - 3y)$ — Use terms in parentheses from Step 1 to fill in the blanks.

Check:

$(2x + 3y)(2x - 3y) = 4x^2 - 6xy + 6xy - 9y^2 = 4x^2 - 9y^2$

b) $81p^2 - 49q^2 = (9p)^2 - (7q)^2$ — $(9p)^2 = 81p^2$. $(7q)^2 = 49q^2$.

$= (\quad + \quad)(\quad - \quad)$ — Write two sets of parentheses.

$= (9p + 7q)(9p - 7q)$ — Use terms in parentheses from Step 1 to fill in the blanks.

Check:

$(9p + 7q)(9p - 7q) = 81p^2 - 63pq + 63pq - 49q^2$
$= 81p^2 - 49q^2$

c) $(a-3b)^2 - 16 = (a-3b)^2 - (4)^2$ — Write each term as a perfect square.

$= (\quad + \quad)(\quad - \quad)$ — Write two sets of parentheses.

$= (a - 3b + 4)(a - 3b - 4)$ — Use terms in parentheses from Step 1 to fill in the blanks.

Factoring perfect square trinomials

A perfect square trinomial can be written as $(x)^2 \pm 2(x)(y) + (y)^2$

perfect square ↑ (under $(x)^2$) perfect square ↑ (under $(y)^2$)

The first and third terms are perfect squares. The middle term is twice the product of the terms in the parentheses. The ± sign stands for plus or minus and means that the middle sign could be plus or minus.

To Factor a Perfect Square Trinomial:

1. Write the trinomial as a Perfect Square Trinomial: $(x)^2 \pm 2(x)(y) + (y)^2$
2. Write a set of parentheses with a square outside. Use the sign of the middle term inside the parentheses: $(\quad \pm \quad)^2$.
3. Fill in the blanks with the perfect squares of the first and third terms. $(x^2 \pm y)^2$

EXAMPLE 32 Factor.

a) $121a^2 - 154a + 49$

b) $25x^2 + 120xy + 144y^2$

c) $(p+5)^2 - 22(p+5) + 121$

d) $x^2 + 6x + 9 - 25y^2$

SOLUTION 32

a) $121a^2 - 154a + 49$

$= (11a)^2 - 2(11a)(7) + (7)^2$ — Write the trinomial as a

	perfect square trinomial.
$= (\quad - \quad)^2$	Write a set of parentheses with a square outside. Use a minus sign inside because the middle term is $-154a$.
$= (11a - 7)^2$	Fill in the blanks from the perfect squares of the first and third terms.
b) $25x^2 + 120xy + 144y^2$	
$= (5x)^2 + 2(5x)(12y) + (12y)^2$	Write the trinomial as a perfect square trinomial.
$= (\quad + \quad)^2$	Write a set of parentheses with a square outside. Use a plus sign inside because the middle term is $+120xy$.
$= (5x + 12y)^2$	Fill in the blanks from the perfect squares of the first and third terms.
c) $(p + 5)^2 - 22(p + 5) + 121$	
Let $y = p + 5$	Make a substitution.
Then	
$(p + 5)^2 - 22(p + 5) + 121 =$ $y^2 - 22y + 121$	Replace $p + 5$ with y.
$y^2 - 22y + 121 = (y - 11)^2$	Factor the new perfect square trinomial.
$[(p + 5) - 11]^2$	Replace y with $p + 5$.
$(p - 6)^2$	Simplify inside the brackets.
d) $x^2 + 6x + 9 - 25y^2$	$x^2 + 6x + 9$ is a perfect square trinomial.
$(x + 3)^2 - 25y^2$	Factor $x^2 + 6x + 9$.
Let $p = x + 3$	Make a substitution.
Then	
$(x + 3)^2 - 25y^2 = p^2 - 25y^2$	Replace $x + 3$ with p.
$p^2 - 25y^2 = (p + 5y)(p - 5y)$	Factor the difference of two squares.
$[(x + 3) + 5y][(x + 3) - 5y]$	Replace p with $x + 3$.

$(x + 3 + 5y)(x + 3 - 5y)$ — Simplify inside the brackets.

Factoring the sum or difference of two cubes

Here is a list of perfect cubes:

$(1)^3 = 1$ $(5)^3 = 125$ $(9)^3 = 729$
$(2)^3 = 8$ $(6)^3 = 216$ $(10)^3 = 1000$
$(3)^3 = 27$ $(7)^3 = 343$
$(4)^3 = 64$ $(8)^3 = 512$

To Factor a Sum or Difference of Two Cubes:

1. Write each term as a perfect cube.
2. Use these formulas to factor: $(x)^3 + (y)^3 = (x + y)(x^2 - xy + y^2)$
   $(x)^3 - (y)^3 = (x - y)(x^2 + xy + y^2)$

You will have to memorize these formulas! There are some similarities about the two formulas that will help you:

1. Each factorization consists of a binomial times a trinomial (two terms times three terms).
2. The terms in the binomial are the cube roots (the numbers in the parentheses) when you write the original terms as perfect cubes.
3. The sign in the binomial is the same sign as in the original problem..
4. The first and third terms of the trinomial are the squares of the terms of the binomial.
5. The middle term of the trinomial is the product of the terms in the binomial, with the opposite sign of the binomial.

EXAMPLE 33 Factor.

a) $8a^3 - 27$

b) $125r^3 + 216s^3$

c) $x^6 - 1$

SOLUTION 33

a) $8a^3 - 27 = (2a)^3 - (3)^3$ — Rewrite each term as a perfect cube.

$= (2a - 3)[(2a)^2 + (2a)(3) + (3)^2]$ — The terms in the binomial are the numbers in parentheses, $2a$ and 3. The first and third terms in the trinomial are the squares of $2a$ and 3. The middle term is the product of $2a$ and 3, with the opposite sign of the binomial, so use a + sign.

$= (2a - 3)(4a^2 + 6a + 9)$ — $(2a)^2 = 4a^2$, $(2a)(3) = 6a$, and $(3)^2 = 9$.

b) $125r^3 + 216s^3 = (5r)^3 + (6s)^3$ — Rewrite each term as a perfect cube.

$= (5r + 6s)[(5r)^2 - (5r)(6s) + (6s)^2]$ — The terms in the binomial are the numbers in parentheses, $5r$ and $6s$. The first and third terms in the trinomial are the squares of $5r$ and $6s$. The middle term is the product of $5r$ and $6s$, with the opposite sign of the binomial, so use a – sign.

$= (5r + 6s)(25r^2 - 30rs + 36s^2)$ — Simplify inside the parentheses.

c) $x^6 - 1$

Note that this can be written as either a difference of two cubes or a difference of two squares. We'll factor it both ways.

As a difference of two cubes:

$x^6 - 1 = (x^2)^3 - (1)^3$ — Rewrite each term as a perfect cube.

$= (x^2 - 1)(x^4 + x^2 + 1)$ — Use the formula for the difference of two cubes.

$= (x - 1)(x + 1)(x^4 + x^2 + 1)$ — Factor $x^2 - 1$ as a difference of two squares.

As a difference of two squares:

$x^6 - 1 = (x^3)^2 - (1)^2$	Rewrite each term as a perfect square.
$= (x^3 + 1)(x^3 - 1)$	Use the formula for the difference of two squares.
$= (x + 1)(x^2 - x + 1)(x - 1)(x^2 + x + 1)$	Factor $x^3 + 1$ as a sum of two cubes and $x^3 - 1$ as a difference of two cubes.

Some textbooks recommend starting with the difference of two squares when there is a choice. Check with your textbook or instructor for their preference.

3.8 A GENERAL APPROACH TO FACTORING

The previous sections have presented various factoring techniques. In our later uses of factoring, problems will not be grouped by type, and therefore you must learn to decide when to do what. It is usually easiest to count the number of terms and follow these steps.

To Factor a Polynomial:

1. Always try to factor out the greatest common factor (GCF) first.
2. If there are 2 terms check for
   a) Difference of two squares: $x^2 - y^2 = (x + y)(x - y)$
   b) Sum of two squares: $x^2 + y^2 = x^2 + y^2$
   c) Difference of two cubes: $x^3 - y^3 = (x - y)(x^2 + xy + y^2)$
   d) Sum of two cubes: $x^3 + y^3 = (x + y)(x^2 - xy + y^2)$
3. If there are 3 terms check for
   a) Perfect square trinomial: $x^2 + 2xy + y^2 = (x + y)^2$
   $x^2 - 2xy + y^2 = (x - y)^2$
   b) Factoring by grouping or by trial and error.
4. If there are 4 or more terms try factoring by grouping.

EXAMPLE 34 Factor.

a) $a^3b^2 - 5a^2b^2$

b) $9x^2 - 81y^2$

c) $250a^3b + 2b^4$

d) $98m^2 + 172n^2$

e) $20a^2y^2 + 100a^2y + 125a^2$

f) $9p^2 - 42pq + 49q^2$

g) $33r^4 - 51r^3 - 30r^2$

h) $2x + 5y + 20xy + 8x^2$

i) $4a^2 - 20ab + 25b^2 - 1$

SOLUTION 34

a) $a^3b^2 - 5a^2b^2 = a^2b^2(a - 5)$

The GCF is a^2b^2.

b) $9x^2 - 81y^2$
$= 9(x^2 - 9y^2)$
$= 9(x + 3y)(x - 3y)$

The GCF is 9. Factor $x^2 - y^2 = (x + y)(x - y)$.

c) $250a^3b + 2b^4$
$= 2b(125a^3 + b^3)$
$= 2b(5a + 6)(25a^2 - 5ab + b^2)$

The GCF is $2b$. Factor $x^3 + y^3 = (x + y)(x^2 - xy + y^2)$

d) $98m^2 + 172n^2 = 2(49m^2 + 81n^2)$

The GCF is 2. Stop here. The sum of two squares does not factor.

e) $20a^2y^2 + 100a^2y + 125a^2$
$= 5a^2(4y^2 + 20y + 25)$
$= 5a^2(2y + 5)^2$

The GCF is $5a^2$. Factor $x^2 + 2xy + y^2 = (x + y)^2$.

f) $9p^2 - 42pq + 49q^2 = (3p - 7q)^2$

Factor $x^2 - 2xy + y^2 = (x - y)^2$.

g) $33r^4 - 51r^3 - 30r^2$
$= 3r^2(11r^2 - 17r - 10)$
$= 3r^2(11r + 5)(r - 2)$

The GCF is $3r^2$. Factor the trinomial by grouping or by trial and error.

h) $2x + 5y + 20xy + 8x^2$

$= (2x + 5y) + (20xy + 8x^2)$ — Try factoring by grouping.

$= 1(2x + 5y) + 4x(5y + 2x)$ — The terms in the parentheses are equal.

$= (1 + 4x)(2x + 5y)$

i) $4a^2 - 20ab + 25b^2 - 1$

$= (4a^2 - 20ab + 25b^2) - 1$ — Try factoring by grouping.

$= (2a - 5b)^2 - 1$ — Factor $x^2 - 2xy + y^2 = (x - y)^2$.

Let $p = 2a - 5b$ so

$= p^2 - 1$ — Replace $2a - 5b$ with p.

$= (p + 1)(p - 1)$ — Factor $x^2 - y^2 = (x + y)(x - y)$.

$= (2a - 5b + 1)(2a - 5b - 1)$ — Replace p with $2a - 5b$.

3.9 SOLVING EQUATIONS BY FACTORING

A **quadratic equation in standard form** is written as

$$ax^2 + bx + c = 0$$

where a, b, and c are real numbers and $a \neq 0$. Many quadratic equations can be solved by factoring and using the zero-factor property:

Zero-Factor Property
If a and b are real numbers and $a \cdot b = 0$, then $a = 0$ or $b = 0$.

Solving factorable quadratic equations

To Solve a Factorable Quadratic Equation:

1. Write the quadratic equation in standard form by getting 0 alone on one side of the equal sign.
2. Factor completely.
3. Set each factor containing a variable equal to 0.
4. Solve each equation.
5. Check your answers in the original equation.

EXAMPLE 35 Solve each equation.

a) $(2x+3)(x-4)=0$

b) $6b^2+11b=-3$

c) $3y^2+21y=0$

d) $\frac{1}{6}a^2-\frac{5}{3}a+4=0$

SOLUTION 35

a) $(2x+3)(x-4)=0$ — This is already completely factored.

$2x+3=0$ or $x-4=0$ — Set each factor equal to 0.

$2x+3-3=0-3$ $\quad x-4+4=0+4$ — Solve each equation.

$2x=-3$ $\quad x=4$

$x=-\frac{3}{2}$ or $x=4$ — Proposed solutions.

Check:

$$\left[2\left(-\frac{3}{2}\right)+3\right]\left[\left(-\frac{3}{2}\right)-4\right]\stackrel{?}{=}0$$

Substitute $-\frac{3}{2}$ into the original equation.

$$[-3+3][-\frac{3}{2}-4]\stackrel{?}{=}0$$

$$0[-\frac{3}{2}-4]=0$$

$-\frac{3}{2}$ checks.

$[2(4)+3][(4)-4] \stackrel{?}{=} 0$ — Substitute 4 into the original equation.

$[2(4)+3][0] \stackrel{?}{=} 0$ — 4 checks.

The solution set is $\{-\frac{3}{2}, 4\}$.

b) $6b^2 + 11b = -3$ — Write the equation in standard form.

$6b^2 + 11b + 3 = 0$ — Add 3 to both sides.

$(3b+1)(2b+3) = 0$ — Factor completely.

$3b + 1 = 0 \quad \text{or} \quad 2b + 3 = 0$ — Set each factor equal to 0.

$3b + 1 - 1 = 0 - 1 \quad 2b + 3 - 3 = 0 - 3$ — Solve each equation.

$3b = -1 \quad 2b = -3$

$b = -\frac{1}{3} \quad \text{or} \quad b = -\frac{3}{2}$ — Proposed solutions. Try the check on your own.

The solution set is $\{-\frac{1}{3}, -\frac{3}{2}\}$.

c) $3y^2 + 21y = 0$ — Equation is in standard form.

$3y(y+7) = 0$ — Factor completely.

$3y = 0 \quad \text{or} \quad y + 7 = 0$ — Set each equation equal to 0.

$y = 0 \quad \text{or} \quad y + 7 - 7 = 0 - 7$ — Solve each equation.

$y = 0 \quad \text{or} \quad y = -7$ — Proposed solutions. Try the check on your own.

The solution set is $\{0, -7\}$.

d) $\frac{1}{6}a^2 - \frac{5}{3}a + 4 = 0$ — The equation is in standard form, but can be factored more easily without fractions.

$6[\frac{1}{6}a^2] - 6 \cdot \frac{5}{3}a + 6 \cdot 4 = 6 \cdot 0$ — Distribute 6.

$a^2 - 10a + 24 = 0$

$(a-4)(a-6) = 0$ — Factor completely.

$a - 4 = 0$ or $a - 6 = 0$ — Set each factor equal to 0.

$a = 4$ or $a = 6$ — Proposed solutions. Try the check on your own.

The solution set is $\{4, 6\}$.

Solving equations with more than two factors

The zero-factor property can be used to solve equations with more than two factors. You must get zero alone on one side of the equal sign. After the polynomial is factored, set each factor containing a variable equal to 0.

EXAMPLE 36 Solve each equation.

a) $x(3x-4)(x+2) = 0$

b) $24s^3 + 4s^2 - 54s - 9 = 0$

SOLUTION 36

a) $x(3x-4)(x+2) = 0$ — The equation is already equal to 0.

$x = 0$ or $3x - 4 = 0$ or $x + 2 = 0$ — Set each factor equal to 0.

$x = 0$ or $x = \frac{4}{3}$ or $x = -2$ — Solve each equation.

Check that the solution set is $\{0, \frac{4}{3}, -2\}$.

b) $24s^3 + 4s^2 - 54s - 9 = 0$ — The equation is already equal to 0.

$(24s^3 + 4s^2) + (-54s - 9) = 0$ — Factor by grouping.

$4s^2(6s+1) - 9(6s+1) = 0$

$(6s+1)(4s^2 - 9) = 0$ — This is not factored completely.

$(6s+1)(2s+3)(2s-3) = 0$ — Factor $4s^2 - 9$ as a

	difference of two squares.
$6s + 1 = 0$ or $2s + 3 = 0$ or $2s - 3 = 0$	Set each factor equal to 0.
$s = -\frac{1}{6}$ or $s = -\frac{3}{2}$ or $s = \frac{3}{2}$	Solve each equation.

Check that the solution set is $\{\frac{1}{6}, -\frac{3}{2}, \frac{3}{2}\}$.

Applications involving factoring

Some word problems lead to quadratic equations that can be solved using the zero-factor property.

EXAMPLE 37 The product of two integers is 63 and the sum of the numbers is 16. Find both integers.

SOLUTION 37

Let x = one of the integers.
Then $16 - x$ = the other integer.

$x(16 - x) = 63$	Write an equation.
$16x - x^2 = 63$	Simplify the left side.
$16x - x^2 - 63 = 0$	Get 0 alone on one side.
$-x^2 + 16x - 63 = 0$	Write the equation in standard form.
$-(x^2 - 16x + 63) = 0$	Factor out -1.
$-(x - 7)(x - 9) = 0$	Factor the trinomial.
$x - 7 = 0$ or $x - 9 = 0$	Set each factor equal to 0.
$x = 7$ or $x = 9$	Solve each equation.

Check:

If $x = 7$	If $x = 9$	
$7(16 - 7) \stackrel{?}{=} 63$	$9(16 - 9) \stackrel{?}{=} 63$	Substitute into the original
$7(9) = 63$	$9(7) = 63$	equation.
When $x = 7$,	When $x = 9$,	
$16 - x = 9$.	$16 - x = 7$.	

Therefore the integers are 7 and 9.

The Pythagorean theorem

Our last application makes use of the Pythagorean theorem:

The Pythagorean theorem:

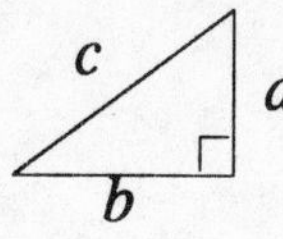

$c^2 = a^2 + b^2$, where c is the length of the hypotenuse and a and b are the lengths of the legs

EXAMPLE 38 The hypotenuse of a right triangle is 25 centimeters. One leg is 5 centimeters longer than the other leg. Find the lengths of both legs.

SOLUTION 38

Let $\quad x =$ the length of one leg.
Then $x + 5 =$ the length of the other leg.

$c^2 = a^2 + b^2$	Use the Pythagorean theorem.
$(25)^2 = (x)^2 + (x+5)^2$	Substitute $c = 25$, $a = x$, and $b = x + 5$.
$625 = x^2 + x^2 + 10x + 25$	Simplify each side.
$625 = 2x^2 + 10x + 25$	Combine similar terms.
$0 = 2x^2 + 10x - 600$	Subtract 625 from both sides.
$2x^2 + 10x - 600 = 0$	Write the equation in standard form.
$2(x^2 + 5x - 300) = 0$	Factor out the GCF, 2.
$2(x - 15)(x + 20) = 0$	Factor $x^2 + 5x - 300$.
$x - 15 = 0$ or $x + 20 = 0$	Set each factor equal to 0 and solve.
$x = 15$ or $x = -20$	Length cannot be negative, so eliminate $x = -20$.

Check:

$25^2 \stackrel{?}{=} (15)^2 + (15+5)^2$	Substitute $x = 15$ into the

original equation.

$625 \stackrel{?}{=} 225 + 400$

$625 = 625$ 15 checks.

If $x = 15$, $x + 5 = 15 + 5 = 20$, so the legs are 15 cm and 20 cm.

Practice Exercises

1. Simplify each expression.

(a) 4^2

(b) $(-4)^2$

(c) -4^2

(d) 6^0

(e) $3x^0, x \neq 0$

(f) 4^{-2}

(g) $7y^{-3}, y \neq 0$

2. Write each number in scientific notation.

(a) 423

(b) 5,280

(c) 0.0475

(d) 0.000368

3. Write each number in expanded form.

(a) 5.32×10^4

(b) 2.16×10^{-1}

(c) 4.37×10^5

(d) 1.23×10^{-3}

4. Simplify. Write all answers with positive exponents.

(a) $3^2 \cdot 3^{-4}$

(b) $x^{-2} \cdot x^{-5} \cdot x^3$

(c) $(2^3)^{-2}$

(d) $(a^{-3})^{-2}$

(e) $(3x)^2$

(f) $(8a^{-2})^{-2}$

5. Simplify. Write all answers with positive exponents.

(a) $\left(\frac{5}{6}\right)^{-2}$

(b) $\left(\frac{3}{x^2}\right)^{-3}$

(c) $\frac{a^{-5}}{a^{-2}}$

(d) $(4p^2q^{-3})^{-2}$

(e) $\frac{6^{-2}m^{-3}n^4}{6^{-1}m^{-5}n^{-3}}$

6. Use scientific notation to compute each of the following.

(a) $(0.00003)^2$

(b) $(5{,}600{,}000)(0.0002)$

(c) $\frac{6,000,000 \cdot 2,700}{0.003 \cdot 900}$

7. Perform the indicated operations.

(a) $(x^2 + xy - y^2) + (2x^2 - 3xy + y^2)$

(b) $(8x^2 - 2xy + y^2) - (-7x^2 - 4xy + 10y^2)$

(c) $(2x^2 - 5xy + y^2) + (-3x^2 - 7xy - 2y^2)$

(d) $(3x^2 - 2x + 5) - (x^2 - 3x - 4)$

8. Perform the indicated operations.

(a) $3x^2(2x^3 - 6x^2 + 5x - 3)$

(b) $(3x - 2y)(4x + 3y)$

(c) $(a - 3b)(3a^2 - 6ab - 5b^2)$

(d) $(2x - 7)(3x + 2)$

(e) $(4p + 3)(2p - 5)$

(f) $(p - q)(p + q)$

(g) $(3r + 2s)(3r - 2s)$

(h) $[(2a + b) - 3][(2a + b) + 3]$

9. Factor out the greatest common factor (GCF).

(a) $6x^3 - 18x^2$

(b) $3a^6 - 4a^5 + a^4$

(c) $24p^2q + 36p^2q^2 - 20pq^2$

(d) $x(x - 5) + y(x - 5)$

(e) $a(2a - 3) - 4(2a - 3)$

(f) $42a^2b^2 + 42$

(g) $-6x^2y^2 - 12x^2y^3 + 15xy^4$

10. Factor by grouping.

(a) $18y^3 + 3y^2 + 30y + 5$

(b) $9xy + 15x + 12y + 20$

(c) $m^2n^2 - 3m^2 - 5n^2 + 15$

(d) $18r^2 + 35s + 30r + 21rs$

11. Factor.

(a) $x^2 + 5x + 6$

(b) $m^2 - 4m - 12$

(c) $a^2 - 6ab + 8b^2$

(d) $x^2 + 5xy - 24y^2$

12. Factor.

(a) $8a^2 - 26ab + 15b^2$

(b) $12m^2 + 17mn + 6n^2$

(c) $6p^2 - 7pq - 5q^2$

(d) $10a^2 + 11ab - 6b^2$

13. Factor.

(a) $2y^3 + 10y^2 + 8y$

(b) $a^3b + 3a^2b - 18ab$

(c) $18m^4 + 15m^3 - 12m^2$

(d) $6x^3y^2 - 25x^2y^3 + 14xy^3$

14. Factor.

(a) $x^4 + 7x^2 + 12$

(b) $3y^4 + 14y^2 + 8$

(c) $10(p + 2)^2 + 7(p + 2) - 12$

(d) $18(p - 3)^2 - 21(p - 3) + 5$

15. Factor.

(a) $25t^2 - 36$

(b) $144x^2 - 1$

(c) $(p - 2q)^2 - 49$

(d) $9a^2 + 42a + 49$

(e) $m^2 - 24m + 144$

(f) $(x - 3)^2 - 18(x - 3) + 81$

(g) $y^2 + 4y + 4 - 25z^2$

16. Factor.

(a) $27p^3 - 125$

(b) $64r^3 + s^3$

(c) $x^6 - 64$

17. Solve.

(a) $(4x + 5)(x - 2) = 0$

(b) $6b^2 + 11b = -4$

(c) $y^2 = 8y$

(d) $\frac{1}{6}x^2 - \frac{1}{2}x - 3$

Answers

1.(a) 16

(b) 16

(c) −16

(d) 1

(e) 3

(f) $\frac{1}{16}$

(g) $\frac{7}{y^3}$

2.

(a) 4.23×10^2

(b) 5.28×10^3

(c) 4.75×10^{-2}

(d) 3.68×10^{-4}

3.

(a) 53200

(b) 0.216

(c) 437,000

(d) 0.00123

4.

(a) $\frac{1}{9}$

(b) $\frac{1}{x^4}$

(c) $\frac{1}{64}$

(d) a^6

(e) $9x^2$

(f) $\frac{a^4}{64}$

5.

(a) $\frac{36}{25}$

(b) $\frac{x^6}{27}$

(c) $\frac{1}{a^3}$

(d) $\frac{q^6}{16p^4}$

(e) $\frac{m^2n^7}{6}$

6.

(a) 9×10^{-10}

(b) 1.12×10^3

(c) 6×10^9

7.

(a) $3x^2 - 2xy$

(b) $15x^2 + 2xy - 9y^2$

(c) $-x^2 - 12xy - y^2$

(d) $2x^2 + x + 9$

8.

(a) $6x^5 - 18x^4 + 15x^3 - 9x^2$

(b) $12x^2 + xy - 6y^2$

(c) $3a^3 - 15a^2b + 13ab^2 + 15b^3$

(d) $6x^2 - 17x - 14$

(e) $8p^2 - 14p - 15$

(f) $p^2 - q^2$

(g) $9r^2 - 4s^2$

(h) $4a^2 + 4ab + b^2 - 9$

9.

(a) $6x^2(x-3)$

(b) $a^4(3a^2-4a+1)$

(c) $4pq(6p+9pq-5q)$

(d) $(x-5)(x+y)$

(e) $(2a-3)(a-4)$

(f) $42(a^2b^2+1)$

(g) $-3xy^2(2x+4xy-5y^2)$ or $3xy^2(-2x-4xy+5y^2)$

10.

(a) $(6y+1)(3y^2+5)$

(b) $(3y+5)(3x+4)$

(c) $(n^2-3)(m^2-5)$

(d) $(3r+5)(6r+7s)$

11.

(a) $(x+2)(x+3)$

(b) $(m-6)(m+2)$

(c) $(a-2b)(a-4b)$

(d) $(x-3y)(x+8y)$

12.

(a) $(4a-3b)(2a-5b)$

(b) $(3m+2n)(4m+3n)$

(c) $(3p-5q)(2p+q)$

(d) $(5a-2b)(2a+3b)$

13.

(a) $2y(y+1)(y+4)$

(b) $ab(a-3)(a+6)$

(c) $3m^2(2m-1)(3m+4)$

(d) $xy^2(2x-7y)(3x-2y)$

14.

(a) $(x^2+4)(x^2+3)$

(b) $(3y^2+2)(y^2+4)$

(c) $(5p+6)(2p+7)$

(d) $(3p-10)(6p-23)$

15.

(a) $(5t-6)(5t+6)$

(b) $(12x-1)(12x+1)$

(c) $(p-2q-7)(p-2q+7)$

(d) $(3a+7)^2$

(e) $(m-12)^2$

(f) $(x-12)^2$

(g) $(y+2-5z)(y+2+5z)$

16.

(a) $(3p-5)(9p^2+15p+25)$

(b) $(4r+s)(16r^2-4rs+s^2)$

(c) $(x+2)(x-2)(x^4-4x^2+16)$

17.

(a) $-\frac{5}{4}, 2$

(b) $-\frac{4}{3}, \frac{1}{2}$

(c) $-3, 6$

(d) $0, -\frac{3}{2}, -6$

4

Rational Expressions

4.1 REDUCING RATIONAL EXPRESSIONS

A rational expression is a fraction that contains polynomials in the numerator and denominator. As with rational numbers, the denominator cannot equal 0.

Restricting variables in the denominator

Because division by 0 is not defined, we must assure that replacing a variable by a number does not make the denominator of a rational expression equal 0.

> **To Find Restrictions on Variables:**
> 1. There are no restrictions if the denominator does not contain a variable.
> 2. If the denominator contains a variable, set it equal to 0 and solve. Use the solutions as restrictions.

EXAMPLE 1 State the restrictions on the variable in the following rational expressions.

a) $\dfrac{4x}{3x-6}$

b) $\dfrac{2x-5}{x^2+4x-21}$

c) $\dfrac{7x-2}{x^2+49}$

SOLUTION 1

a) $\dfrac{4x}{3x-6}$

$3x-6=0$	Set the denominator equal to 0.
$3x=6$	Add 6 to both sides.
$x=2$	Divide both sides by 3.
$x \neq 2$	State the restriction.

b) $\dfrac{2x-5}{x^2+4x-21}$

$x^2+4x-21=0$	Set the denominator equal to 0.
$(x+7)(x-3)=0$	Factor.
$x+7=0$ or $x-3=0$	Set each factor equal to 0.
$x=-7$ or $x=3$	Solve each equation.
$x \neq -7$ and $x \neq 3$	State the restrictions.

c) $\dfrac{7x-2}{x^2+49}$

$x^2+49=0$	Set the denominator equal to 0.

x^2+49 is a sum of two squares and cannot be factored. This denominator can never be 0, so there are no restrictions on x.

Reducing rational expressions

A rational expression is reduced to lowest terms when there are no common factors in the numerator and denominator.

To Reduce a Rational Expression:
1. Factor the numerator and the denominator.
2. Divide out common factors.

EXAMPLE 2 Reduce to lowest terms.

a) $\dfrac{x^2 - 49}{x - 7}$

b) $\dfrac{a^2 + 2a - 15}{a^2 - 9}$

c) $\dfrac{m^2 - mn - 5m + 5n}{m^2 + mn - 5m - 5n}$

SOLUTION 2

a) $\dfrac{x^2 - 49}{x - 7} = \dfrac{(x - 7)(x + 7)}{x - 7}$ Factor the numerator and the denominator.

$= x + 7$ Divide out the common factor $x - 7$.

b) $\dfrac{a^2 + 2a - 15}{a^2 - 9} = \dfrac{(a + 5)(a - 3)}{(a + 3)(a - 3)}$ Factor the numerator and the denominator.

$= \dfrac{a + 5}{a + 3}$ Divide out the common factor $a - 3$.

Note that you cannot divide out a since a is a term (added to something else) rather than a factor (multiplied times something else).

c) $\dfrac{m^2 - mn - 5m + 5n}{m^2 + mn - 5m - 5n} =$ Factor the numerator and denominator by grouping.

$\dfrac{m(m - n) - 5(m - n)}{m(m + n) - 5(m + n)} =$ Finish factoring before dividing out common factors.

$$\frac{(m-n)\,(m-5)}{(m+n)\,(m-5)} =$$

Divide out the common factor $m-5$.

$$\frac{m-n}{m+n}$$

Rational expressions that reduce to –1

An expression like $\frac{b-a}{a-b}$ can be reduced to –1:

$$\frac{b-a}{a-b} = \frac{-a+b}{a-b}$$

Commutative Property: $b-a=-a+b$.

$$= \frac{-1\,(a-b)}{a-b}$$

Factor out –1.

$$=-1$$

Divide out the common factor $a-b$.

To recognize factors that reduce to –1, look for equal variables (or numbers) on opposite sides of subtraction signs:

$$\frac{b-a}{a-b}$$

Equal variables on opposite sides of subtraction signs.

EXAMPLE 3 Write in lowest terms.

a) $\frac{4x-9}{9-4x}$

b) $\frac{x^2-4}{2-x}$

SOLUTION 3

a) $\frac{4x-9}{9-4x} = \frac{-9+4x}{9-4x}$ Use the commutative property.

$= \frac{-1\,(9-4x)}{9-4x}$ Factor out –1.

$= -1$ Divide out the common factor $9-4x$.

b) $\frac{x^2-4}{2-x} = \frac{(x-2)\,(x+2)}{2-x}$ Factor the numerator.

$= \frac{(x-2)\,(x+2)}{-1\,(x-2)}$ Use the commutative property and then factor out –1.

$= \frac{x+2}{-1}$ Divide out the common factor $x-2$.

$= -(x+2)$ or $-x-2$

Warning: Only *factors* can be divided out of the numerator and denominator!

$\frac{a+5}{a+3} \neq \frac{\not{a}+5}{\not{a}+3} \neq \frac{5}{3}$ Here a is a **term** in the numerator and the denominator and *cannot* be cancelled.

$\frac{ab+5b}{ab+3b} = \frac{a\,(b+5)}{a\,(b+3)} = \frac{b+5}{b+3}$ Here a is a **factor** and *can* be divided out.

4.2 MULTIPLYING AND DIVIDING RATIONAL EXPRESSIONS

We can rewrite the multiplication and division fraction rules from Chapter 1 using polynomials as:

Rules for Multiplying and Dividing Rational Expressions:
If A, B, C, and D represent polynomials,

$$\frac{A}{B} \cdot \frac{C}{D} = \frac{AC}{BD} \quad \text{providing } B, D \neq 0$$

$$\frac{A}{B} \div \frac{C}{D} = \frac{A}{B} \cdot \frac{D}{C} = \frac{AD}{BC} \quad \text{providing } B, D \neq 0$$

The key to both multiplication and division of rational expressions is to factor all the numerators and denominators completely.

Multiplying rational expressions

To Multiply Rational Expressions:
1. Factor each numerator and denominator completely.
2. Divide out common factors.
3. Multiply the numerators. Multiply the denominators.
4. Reduce your answer, if possible.

EXAMPLE 4 Multiply.

a) $\dfrac{4x^2}{y^2} \cdot \dfrac{5y^6}{8x}$

b) $\dfrac{2a-2}{b} \cdot \dfrac{6b^3}{4a-4}$

c) $\dfrac{x^2-4}{x^2+5x+6} \cdot \dfrac{x^2+6x+9}{x^2-2x}$

d) $\dfrac{x^2+ax+xy+ay}{x^2+2ax+a^2} \cdot \dfrac{x^2+2xy+y^2}{x^2-ax+xy-ay}$

SOLUTION 4

a) $\dfrac{4x^2}{y^2} \cdot \dfrac{5y^6}{8x} =$

$$\frac{\overset{1\ x}{\cancel{4}\cancel{x^2}}}{\underset{1}{\cancel{y^2}}} \cdot \frac{5\overset{y^4}{\cancel{y^6}}}{\underset{2\ 1}{\cancel{8}\cancel{x}}} =$$ Divide out common factors.

$$\frac{x \cdot 5y^4}{1 \cdot 2} =$$ Multiply the numerators.
Multiply the denominators.

$$\frac{5xy^4}{2}$$

b) $$\frac{2(a-2)}{b} \cdot \frac{6b^3}{4(a-4)} =$$

$$\frac{2a-1}{b} \cdot \frac{6b^3}{4a-1} =$$ Factor.

$$\frac{2(a-1)}{b} \cdot \frac{6b^3}{4(a-1)} =$$ Divide out common factors.

$$\frac{\overset{1}{\cancel{2}}(\overset{1}{\cancel{a-1}})}{\underset{1}{\cancel{b}}} \cdot \frac{\overset{3b^2}{\cancel{6b^3}}}{\underset{\underset{1}{\cancel{2}}}{\cancel{4}}(\underset{1}{\cancel{a-1}})} =$$ Multiply the numerators.
Multiply the denominators.

$$3b^2$$

c) $$\frac{x^2-4}{x^2+5x+6} \cdot \frac{x^2+6x+9}{x^2-2x} =$$

$$\frac{(x-2)\,(x+2)}{(x+2)\,(x+3)} \cdot \frac{(x+3)^2}{x(x-2)} =$$ Factor.

$$\frac{\cancel{(x-2)}\,\cancel{(x+2)}}{\cancel{(x+2)}\,\cancel{(x+3)}} \cdot \frac{\overset{(x+3)}{\cancel{(x+3)^2}}}{x\cancel{(x-2)}} =$$ Divide out common factors.

$$\frac{x+3}{3}$$ Multiply the numerators.
Multiply the denominators.

d) $$\frac{x^2+ax+xy+ay}{x^2+2ax+a^2} \cdot \frac{x^2+2xy+y^2}{x^2-ax+xy-ay} =$$

$$\frac{x(x+a)+y(x+a)}{(x+a)^2} \cdot \frac{(x+y)^2}{x(x-a)+y(x-a)}$$ =Factor.

$$\frac{(x+a)(x+y)}{(x+a^2)} \cdot \frac{(x+y)^2}{(x-a)(x+y)} =$$ Factor completely before dividing out common factors.

$$\frac{\cancel{(x+a)}\,\cancel{(x+y)}}{(x+a)^{\cancel{2}}} \cdot \frac{(x+y)^2}{(x-a)\,\cancel{(x+y)}} =$$ Divide out common factors.

$$\frac{(x+y)^2}{(x+a)(x-a)}$$ Multiply the numerators and denominators.

Dividing rational expressions

To Divide Rational Expressions: Invert the divisor and multiply.

EXAMPLE 5 Divide.

a) $\dfrac{8a^3}{7y^3} \div \dfrac{2a}{49y^9}$

b) $\dfrac{4x^2-1}{6x^2+x-2} \div \dfrac{8x^3+1}{27x^3+8}$

c) $\dfrac{m^4-81}{m^2+m-6} \div \dfrac{m^3+9m}{m^2-4m+4}$

SOLUTION 5

a) $\dfrac{8a^3}{7y^3} \div \dfrac{2a}{49y^9} =$

$\dfrac{8a^3}{7y^3} \bullet \dfrac{49y^9}{2a} =$ Invert the divisor and multiply.

$$\frac{\overset{4a^2}{\cancel{8a^3}}}{\underset{1}{\cancel{7y^3}}} \cdot \frac{\overset{7y^6}{\cancel{49y^9}}}{\underset{1}{\cancel{2a}}} =$$

Divide out common factors.

$$28a^7y^6$$

Multiply the numerators and denominators.

b) $\dfrac{4x^2-1}{6x^2+x-2} \div \dfrac{8x^3+1}{27x^3+8} =$

$$\frac{4x^2-1}{6x^2+x-2} \cdot \frac{27x^3+8}{8x^3+1} =$$

Invert the divisor and multiply.

$$\frac{(2x-1)\,(2x+1)}{(3x+2)\,(2x-1)} \cdot \frac{(3x+2)\,(9x^2-6x+4)}{(2x+1)\,(4x^2-2x+1)} =$$

Factor.

$$\frac{\cancel{(2x-1)}\,\cancel{(2x+1)}}{\cancel{(3x+2)}\,\cancel{(2x-1)}} \cdot \frac{\cancel{(3x+2)}\,(9x^2-6x+4)}{\cancel{(2x+1)}\,(4x^2-2x+1)} =$$

Divide out common factors.

$$\frac{9x^2-6x+4}{4x^2-2x+1}$$

Multiply the numerators and denominators.

c) $\dfrac{m^4-81}{m^2+m-6} \div \dfrac{m^3+9m}{m^2-4m+4} =$

$$\frac{m^4-81}{m^2+m-6} \cdot \frac{m^2-4m+4}{m^3+9m} =$$

Invert the divisor and multiply.

$$\frac{(m^2-9)\,(m^2+9)}{(m+3)\,(m-2)} \cdot \frac{(m-2)^2}{m\,(m^2+9)} =$$

Factor.

$$\frac{(m-3)\,(m+3)\,(m^2+9)}{(m+3)\,(m-2)} \cdot \frac{(m-2)^2}{m\,(m^2+9)} =$$

Keep factoring! m^2-9 is a difference of two squares.

$$\frac{(m-3)\,\cancel{(m+3)}\,\cancel{(m^2+9)}}{\cancel{(m+3)}\,\cancel{(m-2)}} \cdot \frac{(m-2)^{\cancel{2}}}{m\,\cancel{(m^2+9)}} =$$

Divide out common factors.

$$\frac{(m-3)\,(m-2)}{m}$$

Multiply the numerators and denominators.

4.3 ADDING AND SUBTRACTING RATIONAL EXPRESSIONS

It may help you to return to adding and subtracting fractions in Chapter 1 before working with rational expressions.

Adding or subtracting rational expressions with common denominators

To add or subtract rational expressions with a common denominator, add or subtract the numerators and keep the common denominator.

EXAMPLE 6 Add or subtract as indicated.

a) $\dfrac{6}{y+2}+\dfrac{y-3}{y+2}$

b) $\dfrac{x^2+7x-3}{(x+2)\,(x+1)}+\dfrac{2x+17}{(x+2)\,(x+1)}$

c) $\dfrac{4x+3}{2x+7}-\dfrac{3x-8}{2x+7}$

SOLUTION 6

a) $\dfrac{6}{y+2}+\dfrac{y-3}{y+2}=$

The denominators are equal.

$$\frac{6+(y-3)}{y+2}=$$

Add the numerators. Use LCD of $y+2$ as the denominator.

$\dfrac{y+3}{y+2}$ Combine similar terms in the numerator.

b) $\dfrac{x^2+7x-3}{(x+2)(x+1)} + \dfrac{2x+17}{(x+2)(x+1)} =$ The denominators are equal.

$\dfrac{(x^2+7x-3)+(2x+17)}{(x+2)(x+1)} =$ Add the numerators. Use LCD of $(x+2)(x+1)$ as the denominator.

$\dfrac{x^2+9x+14}{(x+2)(x+1)} =$ Combine similar terms in the numerator.

$\dfrac{(x+2)(x+7)}{(x+2)(x+1)} =$ Factor the numerator.

$\dfrac{x+7}{x+1}$ Divide out the common factor $x+2$.

c) $\dfrac{4x+3}{2x+7} - \dfrac{3x-8}{2x+7} =$ The denominators are equal.

$\dfrac{(4x+3)-(3x-8)}{2x+7} =$ Subtract the numerators. Use LCD of $2x+7$ as the denominator.

$\dfrac{4x+3-3x+8}{2x+7} =$ Be careful to distribute the subtraction sign.

$\dfrac{x+11}{2x+7}$ Combine similar terms.

Finding the LCD

To add or subtract rational expressions that do not have a common denominator, we must first find the least common denominator (LCD).

To Find the LCD:

1. Factor each denominator, using exponents where possible.
2. The LCD is the product of each different factor raised to the highest exponent that appears in any of the given expressions.

EXAMPLE 7 Find the LCD for each group of denominators.

a) $5y$, $10y$

b) $x^2 - 4$, $x^2 - 2x - 8$

c) $2x^2 + 4x - 16$, $2x^2 + 12x + 16$

d) $x^2 - 6x + 9$, $x^2 - 2x - 3$

SOLUTION 7

a) $5y = 5 \cdot y$ — Factor each denominator.
$10y = 2 \cdot 5 \cdot y$
$\text{LCD} = 2 \cdot 5 \cdot y = 10y$ — Use each factor in the LCD.

b) $x^2 - 4 = (x - 2)(x + 2)$ — Factor each denominator.
$x^2 - 2x - 8 = (x - 4)(x + 2)$
$\text{LCD} = (x - 2)(x + 2)(x - 4)$ — Use each factor in the LCD.

c) $2x^2 + 4x - 16 = 2(x^2 + 2x - 8)$ — Common factor first.
$= 2(x + 4)(x - 2)$ — Factor the trinomial.
$2x^2 + 12x + 16 = 2(x^2 + 6x + 8)$ — Common factor first.
$= 2(x + 4)(x + 2)$ — Factor the trinomial.
$\text{LCD} = 2(x + 4)(x - 2)(x + 2)$ — Use each factor in the LCD.

d) $x^2 - 6x + 9 = (x - 3)^2$ — Factor each denominator.
$x^2 - 2x - 3 = (x - 3)(x + 1)$
$\text{LCD} = (x - 3)^2(x + 1)$ — Use $(x - 3)^2$ since 2 is the highest exponent that appears on $x - 3$.

Adding or subtracting rational expressions

You must have a common denominator to add or subtract rational expressions. The following rules outline the process.

To Add or Subtract Rational Expressions:

1. Factor each denominator.
2. Find the LCD.
3. Rewrite each rational expression as an equivalent expression with the LCD.
4. Add or subtract numerators. For subtraction, be careful to distribute the subtraction sign across the *whole* numerator following it. The LCD is the denominator.
5. Reduce the answer, if possible, by factoring and dividing out common factors.

EXAMPLE 8 Add or subtract as indicated.

a) $\frac{6}{5y} + \frac{3}{10y}$

b) $\frac{-2}{x^2 - 4} + \frac{3}{x^2 - 2x - 8}$

c) $4 - \frac{5}{3x + 4}$

d) $\frac{4x + 9}{2x^2 + 4x - 16} - \frac{2x + 9}{2x^2 + 12x + 16}$

e) $\frac{2x^2 + 9x - 8}{x - 3} + \frac{x^2 + 2x + 4}{3 - x}$

SOLUTION 8

a) $\frac{6}{5y} + \frac{3}{10y} =$

$\frac{6}{5y} + \frac{3}{2 \cdot 5y} =$ Factor each denominator.

$\frac{2}{2} \cdot \frac{6}{5y} + \frac{3}{2 \cdot 5y} =$ LCD $= 2 \cdot 5y$.

$\frac{12}{10y} + \frac{3}{10y} =$ $\frac{2}{2} \cdot \frac{6}{5y} = \frac{2 \cdot 6}{2 \cdot 5y} = \frac{12}{10y}$

$\frac{15}{10y} =$ Add the numerators. Use LCD of $10y$ as the denominator.

$\frac{3}{2y}$ Reduce answer by dividing out common factor of 5.

b) $\frac{-2}{x^2 - 4} + \frac{3}{x^2 - 2x - 8} =$

$\frac{-2}{(x-2)\,(x+2)} + \frac{3}{(x-4)\,(x+2)} =$ Factor each denominator.

$\frac{x-4}{x-4} \cdot \frac{-2}{(x-2)\,(x+2)} + \frac{x-2}{x-2} \cdot \frac{3}{(x-4)\,(x+2)} =$ The LCD is $(x-4)(x-2)(x+2)$.

$\frac{-2x+8}{(x-4)\,(x-2)\,(x+2)} + \frac{3x-6}{(x-4)\,(x+2)\,(x-2)} =$ Multiply out each numerator. Leave denominators factored.

$\frac{(-2x+8) + (3x-6)}{(x-4)\,(x-2)\,(x+2)} =$ Add the numerators. Use LCD as the denominator.

$\frac{x+2}{(x-4)\,(x-2)\,(x+2)} =$ Combine similar terms.

$\frac{1}{(x-4)\,(x-2)}$ Reduce answer by dividing out common factor of $x + 2$.

c) $4 - \frac{5}{3x+4} =$

$\frac{4}{1} - \frac{5}{3x+4} =$ The LCD is $3x + 4$.

$\frac{3x+4}{3x+4} \cdot \frac{4}{1} - \frac{5}{3x+4} =$

$$\frac{12x+16}{3x+4} - \frac{5}{3x+4} =$$

Multiply out the numerators.

$$\frac{12x+16-5}{3x+4} =$$

Subtract the numerators. Use the LCD as the denominator.

$$\frac{12x+11}{3x+4}$$

d) $$\frac{4x+9}{2x^2+4x-16} - \frac{2x+9}{2x^2+12x+16} =$$

$$\frac{4x+19}{2(x+4)(x-2)} - \frac{2x+9}{2(x+2)(x+4)} =$$

Factor each denominator.

$$\frac{x+2}{x+2} \bullet \frac{4x+19}{2(x+4)(x-2)} - \frac{x-2}{x-2} \bullet \frac{2x+9}{2(x+2)(x+4)} =$$

The LCD is $2(x+2)(x-2)(x+4)$.

$$\frac{4x^2+27x+38}{2(x+2)(x+4)(x-2)} - \frac{2x^2+5x-18}{2(x+2)(x+4)(x-2)} =$$

Multiply out the numerators. Leave denominators factored.

$$\frac{(4x^2+27x+38)-(2x^2+5x-18)}{2(x+2)(x+4)(x-2)} =$$

Subtract the numerators. Use LCD as denominator.

$$\frac{2x^2+22x+56}{2(x+2)(x+4)(x-2)} =$$

Now factor the numerator.

$$\frac{2(x^2+11x+28)}{2(x+2)(x+4)(x-2)} =$$

$$\frac{2(x+4)(x+7)}{2(x+2)(x+4)(x-2)} =$$

Divide out common factors.

$$\frac{x+7}{(x+2)(x-2)}$$

e) $\dfrac{2x^2+9x-8}{x-3}+\dfrac{x^2+2x+4}{3-x}=$

These denominators are opposites.

$\dfrac{2x^2+9x-8}{x-3}+\dfrac{-1}{-1}\bullet\dfrac{x^2+2x+4}{3-x}=$

Multiply by $\frac{-1}{-1}$ to get the LCD.

$\dfrac{2x^2+9x-8}{x-3}+\dfrac{-x^2-2x-4}{x-3}=$

Add the numerators. Use the LCD as the denominator.

$\dfrac{x^2+7x-12}{x-3}=$

Combine similar terms in the numerator.

$\dfrac{(x-4)\,(x-3)}{x-3}=$

Factor the numerator.

$x-4$

Divide out the common factor $x-3$.

4.4COMPLEX FRACTIONS

A complex fraction is a rational expression that contains fractions in the numerator and/or denominator. There are two methods used to simplify complex fractions, each of which will be discussed in this section.

Simplifying complex fractions using method 1

In method 1, we need to write the numerator and denominator of the complex fraction as single fractions.

To Simplify a Complex Fraction Using Method 1:

1. Simplify the numerator of the complex fraction to a single fraction.
2. Simplify the denominator of the complex fraction to a single fraction.
3. Perform the indicated division by inverting the divisor and multiplying.
4. Reduce, if possible.

EXAMPLE 9 Simplify each complex fraction using Method 1.

a) $\dfrac{\dfrac{4x+8}{3x^2}}{\dfrac{4}{6x}}$

b) $\dfrac{\dfrac{3x+1}{x^2-49}}{\dfrac{9x^2-1}{x-7}}$

c) $\dfrac{4+\dfrac{3}{x}}{2-\dfrac{1}{x}}$

SOLUTION 9

a) $\dfrac{\dfrac{4x+8}{3x^2}}{\dfrac{4}{6x}} =$ The numerator and denominator are already single fractions.

$\dfrac{4x+8}{3x^2} \cdot \dfrac{6x}{4} =$ Invert and multiply.

$\dfrac{\overset{1}{\cancel{4}}(x+2)}{\underset{1x}{\cancel{3}x^{\cancel{2}}}} \cdot \dfrac{\overset{2}{\cancel{6x}}}{\underset{1}{\cancel{4}}} =$ Factor and divide out common factors.

$$\frac{2(x+2)}{x}$$

b) $$\frac{\frac{3x+1}{x^2-49}}{\frac{9x^2-1}{x-7}} =$$

The numerator and denominator are already single fractions.

$$\frac{3x+1}{x^2-49} \cdot \frac{x-7}{9x^2-1} =$$

Invert and multiply.

$$\frac{\cancel{3x+1}}{(\cancel{x-7})(x+7)} \cdot \frac{\cancel{x-7}}{(3x-1)(\cancel{3x+1})} =$$

Factor completely and divide out common factors.

$$\frac{1}{(x+7)(3x-1)}$$

c) $$\frac{4+\frac{3}{x}}{2-\frac{1}{x}} =$$

$$\frac{\frac{x}{x}\cdot\frac{4}{1}+\frac{3}{x}}{\frac{x}{x}\cdot\frac{2}{1}-\frac{1}{x}} =$$

Simplify the numerator and denominator.

$$\frac{\frac{4x+3}{x}}{\frac{2x-1}{x}} =$$

Add the fractions in the numerator, subtract the fractions in the denominator.

$$\frac{4x+3}{x} \cdot \frac{x}{2x-1}$$

Invert and multiply.

$$\frac{4x+3}{2x-1}$$

Divide out the common factor x.

Simplifying complex fractions using Method 2

Method 2, sometimes called the LCD method, uses the LCD of the denominators within the complex fraction and the distributive property to simplify the complex fraction.

To Simplify Complex Fractions Using Method 2:

1. Find the LCD of *all* the denominators within the complex fraction.
2. Multiply the numerator and denominator of the complex fraction by the LCD.
3. Simplify and reduce, if possible.

EXAMPLE 10 Simplify each complex fraction using Method 2.

a) $$\frac{4+\frac{3}{x}}{2-\frac{1}{x}}$$

b) $$\frac{2+\frac{5}{x}-\frac{3}{x^2}}{2-\frac{5}{x}+\frac{2}{x^2}}$$

c) $$\frac{a^{-1}+b^{-1}}{a^{-1}-b^{-1}}$$

SOLUTION 10

a) $$\frac{4+\frac{3}{x}}{2-\frac{1}{x}} =$$

$$\frac{x(4+\frac{3}{x})}{x(2-\frac{1}{x})} =$$ Multiply the numerator and denominator by the LCD.

$$\frac{x(4)+x(\frac{3}{x})}{x(2)-x(\frac{1}{x})}=$$

Use the distributive property to multiply the LCD times each term.

$$\frac{4x+3}{2x-1}$$

b) $$\frac{2+\frac{5}{x}-\frac{3}{x^2}}{2-\frac{5}{x}+\frac{2}{x^2}}=$$

$$\frac{x^2(2+\frac{5}{x}-\frac{3}{x^2})}{x^2(2-\frac{5}{x}+\frac{2}{x^2})}=$$

Multiply the numerator and denominator by the LCD.

$$\frac{x^2(2)+x^2(\frac{5}{x})-x^2(\frac{3}{x^2})}{x^2(2)-x^2(\frac{5}{x})+x^2(\frac{2}{x^2})}=$$

Use the distributive property to multiply the LCD times each term.

$$\frac{2x^2+5x-3}{2x^2-5x+2}=$$

Simplify.

$$\frac{(2x-1)(x+3)}{(2x-1)(x-2)}=$$

Factor the numerator and denominator.

$$\frac{x+3}{x-2}$$

Divide out the common factor $2x-1$.

c) $$\frac{a^{-1}+b^{-1}}{a^{-1}-b^{-1}}=$$

$$\frac{\frac{1}{a}+\frac{1}{b}}{\frac{1}{a}-\frac{1}{b}}=$$

Use the definition of negative exponents to

rewrite $a^{-1} = \frac{1}{a}$ and $b^{-1} = \frac{1}{b}$.

$$\frac{ab\left(\frac{1}{a} + \frac{1}{b}\right)}{ab\left(\frac{1}{a} - \frac{1}{b}\right)} =$$

Multiply the numerator and denominator by the LCD.

$$\frac{ab\left(\frac{1}{a}\right) + ab\left(\frac{1}{b}\right)}{ab\left(\frac{1}{a}\right) - ab\left(\frac{1}{b}\right)} =$$

Use the distributive property to multiply the LCD times each term.

$$\frac{b+a}{b-a}$$

Simplify.

4.5 DIVIDING POLYNOMIALS

We added, subtracted, and multiplied polynomials in Chapter 3. In this section we will divide polynomials. We will begin by dividing by monomials (one term polynomials).

Dividing a polynomial by a monomial

In a division problem like $6 \div 2 = 3$, 6 is called the dividend, 2 is the divisor and 3 is the quotient. When the divisor is a monomial, we can use the following rules.

To Divide a Polynomial by a Monomial:
Divide each term of the dividend by the monomial by:

1. Dividing the numerical coefficients. Leave as fractions if the numbers do not divide evenly.
2. Subtracting exponents on equal bases. Write the answer with positive exponents.

EXAMPLE 11 Divide.

a) $\dfrac{14x^6 - 28x^5 + 21x^2}{7x^2}$

b) $\dfrac{8a^3b^3 - 9a^2b - 4b^2}{2ab^2}$

SOLUTION 11

a) $\dfrac{14x^6 - 28x^5 + 21x^2}{7x^2} =$

$\dfrac{14x^6}{7x^2} - \dfrac{28x^5}{7x^2} + \dfrac{21x^2}{7x^2} =$ Divide each term by $7x^2$.

$2x^4 - 4x^3 + 3$ Divide coefficients, subtract exponents.

b) $\dfrac{8a^3b^3 - 9a^2b - 4b^2}{2ab^2} =$

$\dfrac{8a^3b^3}{2ab^2} - \dfrac{9a^2b}{2ab^2} - \dfrac{4b^2}{2ab^2} =$ Divide each term by $2ab^2$.

$4a^2b - \dfrac{9a}{2b} - \dfrac{2}{a}$ Write the answer with positive exponents.

Dividing a polynomial by a polynomial

Dividing a polynomial by a polynomial is very similar to long division with whole numbers. The process is outlined below.

Dividing a Polynomial by a Polynomial:

1. Arrange the dividend and the divisor in descending order. Insert 0's for missing powers.
2. Divide the first term of the dividend by the first term of the divisor. Write this quotient on the line above the dividend.
3. Multiply the quotient times the divisor. Write this product below the dividend. Subtract and bring down the next term.
4. This becomes the new dividend. Go back to Step 2 using the new dividend.
5. Check by multiplying the divisor times the quotient, and adding the remainder. This product should equal the dividend.

EXAMPLE 12 Divide.

a) $\dfrac{3x^2 - 7x - 26}{x - 4}$

b) $3x - 2\overline{)9x^3 + 11x - 2}$

c) $\dfrac{p^3 + 64}{p + 4}$

SOLUTION 12

a) $\dfrac{3x^2 - 7x - 26}{x - 4}$

$$\begin{array}{r} 3x \\ x-4\overline{)3x^2 - 7x - 26} \\ \underline{-(3x^2 - 12x)} \\ 5x - 26 \end{array}$$

Divide $\dfrac{3x^2}{x} = 3x$

Multiply $3x(x-4) = 3x^2 - 12x$

Subtract and bring down -26

$$\begin{array}{r} 3x+5 \\ x-4\overline{)3x^2-7x-26} \\ \underline{-(3x^2-12x)} \\ 5x-26 \\ \underline{-(5x-20)} \\ -6 \end{array}$$

Divide $\dfrac{5x}{x}=5$

Multiply $5(x-4)=5x-20$

Subtract

The remainder, –6, is usually written as $\dfrac{-6}{x-4}$.

Thus,

$$\frac{3x^2-7x-26}{x-4} = 3x+5+\frac{-6}{x-4}.$$

Check:

$$\begin{array}{r} (x-4)(3x+5) = 3x^2-7x-20 \\ \underline{+-6} \\ 3x^2-7x-26 \end{array}$$

Multiply the divisor times the quotient. Add the remainder.

This is the dividend.

b) $3x-2\overline{)9x^3+0x^2+11x-2}$

Insert $0x^2$ for the missing term in the dividend.

$$\begin{array}{r} 3x^2 \\ 3x-2\overline{)9x^3+0x^2+11x-2} \\ \underline{-(9x^3-6x^2)} \\ 6x^2+11x \end{array}$$

Divide $\dfrac{9x^3}{3x}=3x^2$

Multiply $3x^2(3x-2)=9x^3-6x^2$

Subtract and bring down $11x$

Thus $\dfrac{9x^3+11x-2}{3x-2} = (3x^2+2x+5)+\dfrac{8}{3x-2}$.

Try the check on your own by multiplying $(3x-2)(3x^2+2x+5)$ and adding the remainder 8 to your product.

c)

$$p+4\overline{)p^3+0p^2+0p+64}$$

Insert $0p^2$ and $0p$ for missing terms in the dividend.

$$\begin{array}{r}
p^2 \\
p+4\overline{)p^3+0p^2+0p+64} \\
\underline{-(p^3+4p^2)} \\
-4p^2+0p
\end{array}$$

Divide $\dfrac{p^3}{p}=p^2$

Multiply $p^2(p+4)=p^3+4p^2$

Subtract and bring down $+0p$

$$\begin{array}{r}
p^2-4p \\
p+4\overline{)p^3+0p^2+0p+64} \\
\underline{-(p^3+4p^2)} \\
-4p^2+0p \\
\underline{-(-4p^2-16p)} \\
16p+64
\end{array}$$

Divide $\dfrac{-4p^2}{p}=-4p$

Multiply $-4p(p+4)=-4p^2-16p$

Subtract and bring down $+64$

$$\begin{array}{r}
p^2-4p+16 \\
p+4\overline{)p^3+0p^2+0p+64} \\
\underline{-(p^3+4p^2)} \\
-4p^2+0p \\
\underline{-(-4p^2-16p)} \\
16p+64 \\
\underline{-(16p+64)} \\
0
\end{array}$$

Divide $\dfrac{16p}{p}=16$

Multiply $16(p+4)=16p+64$

Subtract

4.6 SOLVING EQUATIONS COMBINING RATIONAL EXPRESSIONS

When an equation contains one or more rational expressions, it is usually easiest to clear the fractions by multiplying both sides of the equation by the LCD.

To Solve Equations Containing Rational Expressions:

1. Factor each denominator as necessary to find the LCD.
2. Multiply both sides of the equation by the LCD.
3. Use the distributive property to multiply each term by the LCD.
4. Divide out common factors. There should be *no* fractions remaining after this step.
5. Solve the resulting equation.
6. Check each solution in the original equation.

In step 5, the resulting equation may be a linear equation (get the variable on one side, number on the other) or a quadratic equation (get 0 alone on one side of the equation).

Solving rational equations whose solutions check

EXAMPLE 13 Solve each equation and check your solutions.

a) $\frac{x}{4} - \frac{2x}{5} = 3$

b) $\frac{2}{a+5} = \frac{2}{5} - \frac{a}{a+5}$

c) $1 - \frac{1}{x} = \frac{12}{x^2}$

d) $\frac{3}{x^2-1} - \frac{1}{x^2+3x-4} = \frac{1}{x^2+5x+4}$

SOLUTION 13

a) $\frac{x}{4} - \frac{2x}{5} = 3$

$20\left(\frac{x}{4} - \frac{2x}{5}\right) = 20(3)$ Multiply both sides of the equation by the LCD, 20.

$20\left(\frac{x}{4}\right) - 20\left(\frac{2x}{5}\right) = 60$ Use the distributive property.

$5x - 8x = 60$ Divide out common factors.

$-3x = 60$ Combine similar terms.

$\frac{-3x}{-3} = \frac{60}{-3}$ Divide both sides by –3.

$x = -20$ Proposed solution.

Check:

$\frac{-20}{4} - \frac{2(-20)}{5} \stackrel{?}{=} 3$ Substitute $x = -20$ in the original equation.

$-5 + 8 \stackrel{?}{=} 3$

$3 = 3$ The solution checks.

$\{-20\}$ is the solution set.

b) $\frac{2}{a+5} = \frac{2}{5} - \frac{a}{a+5}$

$5(a+5)\left(\frac{2}{a+5}\right) = 5(a+5)\left(\frac{2}{5} - \frac{a}{a+5}\right)$ Multiply both sides by the LCD, $5(a + 5)$.

$5(a+5)\left(\frac{2}{a+5}\right) = 5(a+5)\left(\frac{2}{5}\right) - 5(a+5)\left(\frac{a}{a+5}\right)$

Use the distributive property.

$5(2) = (a + 5)(2) - 5(a)$	Divide out common factors.
$10 = 2a + 10 - 5a$	Simplify each side.
$10 = -3a + 10$	Combine similar terms.
$10 - 10 = -3a + 10 - 10$	Subtract 10 from both sides.
$0 = -3a$	Combine similar terms.
$\frac{0}{-3} = \frac{a}{-3}$	Divide by –3.
$0 = a$	Proposed solution.

Check:

$\frac{2}{(0) + 5} \stackrel{?}{=} \frac{2}{5} - \frac{(0)}{(0) + 5}$	Substitute 0 for a in the original equation.
$\frac{2}{5} \stackrel{?}{=} \frac{2}{5} - \frac{0}{5}$	
$\frac{2}{5} \stackrel{?}{=} \frac{2}{5} - 0$	
$\frac{2}{5} = \frac{2}{5}$	True statement.

{0} is the solution set.

c) $1 - \frac{1}{x} = \frac{12}{x^2}$

$x^2(1 - \frac{1}{x}) = x^2(\frac{12}{x^2})$	Multiply both sides by the LCD, x^2.
$x^2(1) - x^2(\frac{1}{x}) = x^2(\frac{12}{x^2})$	Use the distributive property.
$x^2 - x = 12$	Divide out common factors.
$x^2 - x - 12 = 0$	Get 0 alone on one side by subtracting 12 from both sides.

$(x-4)(x+3)=0$ Factor.

$x-4=0$ or $x+3=0$ Set each factor equal to 0.

$x=4$ or $x=-3$ Solve each equation.

Check:

when $x=4$

$$1-\frac{1}{4}=\frac{12}{4^2}$$

$$\frac{4}{4}-\frac{1}{4}\stackrel{?}{=}\frac{12}{16}$$

$$\frac{3}{4}=\frac{3}{4}$$

when $x=-3$

$$1-\frac{1}{(-3)}\stackrel{?}{=}\frac{12}{(-3)^2}$$

$$\frac{3}{3}+\frac{1}{3}\stackrel{?}{=}\frac{12}{9}$$

$$\frac{4}{3}=\frac{4}{3}$$

Both solutions check.

$\{-3, 4\}$ is the solution set.

d) $$\frac{3}{x^2-1}-\frac{1}{x^2+3x-4}=\frac{1}{x^2+5x+4}$$

$$\frac{3}{(x-1)(x+1)}-\frac{1}{(x+4)(x-1)}=\frac{1}{(x+4)(x+1)}$$

Factor each denominator.

$$(x-1)(x+1)(x+4)\left(\frac{3}{(x-1)(x+1)}-\frac{1}{(x+4)(x-1)}\right)=$$

$$(x-1)(x+1)(x+4)\left(\frac{1}{(x+4)(x+1)}\right)$$

Multiply both sides by the LCD.

$$(x-1)(x+1)(x+4)\left(\frac{3}{(x-1)(x+1)}\right)-(x-1)(x+1)(x+4)\left(\frac{1}{(x+4)(x-1)}\right)=$$

$$(x-1)(x+1)(x+4)\left(\frac{1}{(x+4)(x+1)}\right)$$

Use the distributive property.

$(x+4)(3)-(x+1)(1)=(x-1)(1)$ Divide out common

factors.

$$3x + 12 - x - 1 = x - 1$$ Use the distributive property.

$$2x + 11 = x - 1$$ Combine similar terms.

$$2x + 11 - x = x - 1 - x$$ Subtract x from both sides.

$$x + 11 = -1$$ Combine similar terms.

$$x + 11 - 11 = -1 - 11$$ Subtract 11 from both sides.

$$x = -12$$

Check:

$$\frac{3}{(-12)^2 - 1} - \frac{1}{(-12)^2 + 3(-12) - 4} \stackrel{?}{=} \frac{1}{(-12)^2 + 5(-12) + 4}$$

Substitute -12 for x in the original equation.

$$\frac{3}{143} - \frac{1}{104} \stackrel{?}{=} \frac{1}{88}$$

$$\frac{169}{(143)(104)} \stackrel{?}{=} \frac{1}{88}$$

$$\frac{1}{88} = \frac{1}{88}$$

The solution set is $\{-12\}$.

The rational equations we've solved had solutions that checked. Recall that division by 0 is not defined, and when a proposed solution makes a denominator 0, that proposed solution cannot be included in the solution set.

EXAMPLE 14 Solve and check.

a) $\frac{2y}{y-3} + \frac{4}{3} = \frac{6}{y-3}$

b) $\frac{p+3}{p^2 - p} = \frac{8}{p^2 - 1}$

SOLUTION 14

a) $\frac{2y}{y-3} + \frac{4}{3} = \frac{6}{y-3}$

$3(y-3)\left(\frac{2y}{y-3} + \frac{4}{3}\right) = 3(y-3)\left(\frac{6}{y-3}\right)$ Multiply both sides by the LCD, $3(y-3)$.

$3(y-3)\left(\frac{2y}{y-3}\right) + 3(y-3)\left(\frac{4}{3}\right) = 3(y-3)\left(\frac{6}{y-3}\right)$

Use the distributive property.

$3(2y) + (y-3)(4) = 3(6)$ Divide out common factors.

$6y + 4y - 12 = 18$ Multiply.

$10y - 12 = 18$ Add similar terms.

$10y - 12 + 12 = 18 + 12$ Add 12 to both sides.

$10y = 30$

$\frac{10y}{10} = \frac{30}{10}$ Divide both sides by 10.

$y = 3$ Proposed solution.

Check:

$\frac{2(3)}{3-3} + \frac{4}{3} \stackrel{?}{=} \frac{6}{3-3}$ Substitute 3 for y in the original equation.

$\frac{6}{0} + \frac{4}{3} \stackrel{?}{=} \frac{6}{0}$

Since division by 0 is not defined, the solution set is the empty set, $\varnothing$.

b) $\frac{p+3}{p^2-p} = \frac{8}{p^2-1}$

$\frac{p+3}{p(p-1)} = \frac{8}{(p-1)(p+1)}$ Factor each denominator.

$p(p-1)(p+1)\left(\frac{p+3}{p(p-1)}\right) = p(p-1)(p+1)\left(\frac{8}{(p-1)(p+1)}\right)$

Multiply both sides by the LCD.

$(p+1)(p+3) = p(8)$ Divide out common factors.

$p^2 + 4p + 3 = 8p$ Multiply.

$p^2 + 4p + 3 - 8p = 8p - 8p$ Subtract $8p$ from both sides.

$p^2 - 4p + 3 = 0$ Combine similar terms.

$(p-3)(p-1) = 0$ Factor.

$p - 3 = 0$ or $p - 1 = 0$ Set each factor equal to 0.

$p = 3$ or $p = 1$ Solve each equation.

Check:

If $p = 3$

$\frac{3+3}{(3)^2-3} \stackrel{?}{=} \frac{8}{(3)^2-1}$

$\frac{6}{6} \stackrel{?}{=} \frac{8}{8}$

$1 = 1$

If $p = 1$

$\frac{1+3}{(1)^2-1} \stackrel{?}{=} \frac{8}{(1)^2-1}$

$\frac{4}{0} \stackrel{?}{=} \frac{8}{0}$

$p = 1$ does not check

Since $p = 1$ makes the denominators equal zero, the solution set is $\{3\}$.

4.7 APPLICATIONS

When an equation contains rational expressions, we multiply both sides of the equation by the LCD. The formulas and applications in this section will involve equations containing rational expressions.

Formulas

If your calculator has a list of formulas with it, the following formulas may appear on your list. Formulas containing rational expressions appear in almost every area of study.

EXAMPLE 15 Solve each formula for the specified variable.

a) $F = \frac{mv^2}{r}$ for m (centrifugal force)

b) $\frac{1}{f} = \frac{1}{p} + \frac{1}{q}$ for q (focal length of a lens)

SOLUTION 15

a) $F = \frac{mv^2}{r}$ for m

$r(F) = r\left(\frac{mv^2}{r}\right)$ Multiply both sides by the LCD, r.

$rF = mv^2$ Divide out common factors.

$\frac{rF}{v^2} = \frac{mv^2}{v^2}$ Divide both sides by v^2.

$\frac{rF}{v^2} = m$ Formula is solved for m.

b) $\frac{1}{f} = \frac{1}{p} + \frac{1}{q}$ for q

$fpq\left(\frac{1}{f}\right) = fpq\left(\frac{1}{p} + \frac{1}{q}\right)$ Multiply both sides by the LCD.

$fpq\left(\frac{1}{f}\right) = fpq\left(\frac{1}{p}\right) + fpq\left(\frac{1}{q}\right)$	Use the distributive property.
$pq = fq + fp$	Divide out common factors.
$pq - fq = fq + fp - fq$	Get terms containing q on one side.
$pq - fq = fp$	Combine similar terms.
$q(p - f) = fp$	Factor out q.
$\frac{p(p-f)}{p-f} = \frac{fp}{p-f}$	Divide both sides by $p - f$.
$q = \frac{fp}{p-f}$	Formula is now solved for q.

Number applications

EXAMPLE 16 If the numerator of the fraction $\frac{2}{5}$ is multiplied by a number and four times that number is added to the denominator, the resulting fraction equals $\frac{6}{17}$. What is the number?

SOLUTION 16

Let x = the number

Then the numerator, 2, multiplied by a number = $2x$.

Four times that number added to the denominator, 5, = $4x + 5$

So

$\frac{2x}{4x+5} = \frac{6}{17}$	Write an equation.
$17(4x+5)\left(\frac{2x}{4x+5}\right) = 7(4x+5)\left(\frac{6}{17}\right)$	Multiply both sides by the LCD.
$17(2x) = (4x + 5)(6)$	Divide out common factors.
$34x = 24x + 30$	Multiply.
$34x - 24x = 24x + 30 - 24x$	Subtract $24x$ from both

sides.

$10x = 30$ — Combine similar terms.

$\frac{10x}{10} = \frac{30}{10}$ — Divide both sides by 10.

$x = 3$ — Proposed solution.

Check:

$\frac{2(3)}{4(3)+5} \stackrel{?}{=} \frac{6}{17}$ — Substitute $x = 3$ into the original equation.

$\frac{6}{17} \stackrel{?}{=} \frac{6}{17}$ — True.

The number is 3.

Work applications

Work problems involve the length of time to do a job and usually involve two or more people or machines working at a task such as painting or typing. These techniques are also used for pipes that fill and drain sinks or swimming pools. Study the following table that relates total job time to the fractional part of the job completed in 1 hour.

Time to Complete Job	Part of Job Completed in 1 Hour	Part of Job Completed in 2 Hours
5 hours	$\frac{1}{5}$	$\frac{2}{5}$
8 hours	$\frac{1}{8}$	$\frac{2}{8}$
x hours	$\frac{1}{x}$	$\frac{2}{x}$

EXAMPLE 17 A dot matrix printer can complete a job in 2 hours. A laser printer can complete the same job in 5 hours. If both printers are used to complete the job, how long will it take to complete the job?

SOLUTION 17

Let x = the time to complete the job with both printers

In 1 hour, the dot matrix printer completes $\frac{1}{2}$ of the job.

In 1 hour, the laser printer completes $\frac{1}{5}$ of the job.

Together, the printers complete $\frac{1}{x}$ of the job.
So,

$$\frac{1}{2} + \frac{1}{5} = \frac{1}{x}$$ Write an equation.

$$10x\left(\frac{1}{2} + \frac{1}{5}\right) = 10x\left(\frac{1}{x}\right)$$ Multiply both sides by the LCD.

$$10x\left(\frac{1}{2}\right) + 10x\left(\frac{1}{5}\right) = 10x\left(\frac{1}{x}\right)$$ Use the distributive property.

$$5x + 2x = 10$$ Divide out common factors.

$$7x = 10$$ Add similar terms.

$$\frac{7x}{7} = \frac{10}{7}$$ Divide both sides by 7.

$$x = \frac{10}{7}$$

It takes $\frac{10}{7}$ hours for the printers to complete the job.

EXAMPLE 18 A sink can be filled with water in 5 minutes. It takes 12 minutes to empty the sink when the drain is open. If the drain is accidently left open, how long will it take to fill the sink?

SOLUTION 18

Let x = the time to fill the sink.

In 1 minute, the faucet fills $\frac{1}{5}$ of the sink.

In 1 minute, the drain empties $\frac{1}{12}$ of the sink.

In 1 minute, $\frac{1}{x}$ of the sink is filled.
So,

$$\frac{1}{5} - \frac{1}{12} = \frac{1}{x}$$ Write an equation.

$$60x\left(\frac{1}{5} - \frac{1}{12}\right) = 60\left(\frac{1}{x}\right)$$ Multiply both sides by the LCD.

$$60x\left(\frac{1}{5}\right) - 60x\left(\frac{1}{12}\right) = 60x\left(\frac{1}{x}\right)$$ Use the distributive property.

$$12x - 5x = 60$$ Divide out common factors.

$$7x = 60$$ Combine similar terms.

$$\frac{7x}{7} = \frac{60}{7}$$ Divide both sides by 7.

$$x = \frac{60}{7}$$

It takes $\frac{60}{7}$ hours ($8\frac{4}{7}$ hours) to drain the sink.

D = RT applications

Some applications leading to equations containing rational expressions use the formula

$D = RT$ Distance = Rate times Time

A table is a useful way to organize data for this type of problem, as demonstrated in the next example.

EXAMPLE 19 A biathlon contains a 20K (20 kilometers) bike ride and a 6K run. If Joan can bike $2\frac{1}{2}$ times as fast as she runs, and her 6K run takes 30 minutes longer than her 20K bike ride, find her biking rate and her running rate.

SOLUTION 19

	R	T	D
Biking	$2.5x$		20
Running	x		6

Let's begin by filling in the rate and distance columns.

Let x = Joan's running rate

Then $2.5x$ = Joan's biking rate

Since

$$RT = D$$ Distance formula.

$$\frac{RT}{R} = \frac{D}{R}$$ Divide both sides by R.

$$T = \frac{D}{R}$$ Distance formula solved for T.

We can now fill in the T column:

	R	T	D
Biking	$2.5x$	$\frac{20}{2.5x}$	20
Running	x	$\frac{6}{x}$	6

Since the run takes 30 minutes longer ($\frac{1}{2}$ hour) than the bike ride,

$$\frac{20}{2.5x} = \frac{6}{x} + \frac{1}{2}$$ Write an equation.

$$2(2.5x)\left(\frac{20}{2.5x}\right) = 2(2.5x)\left(\frac{6}{x} + \frac{1}{2}\right)$$

Leave the LCD in factored form.

$$2(2.5x)\left(\frac{20}{2.5x}\right) = 2(2.5x)\left(\frac{6}{x}\right) + 2(2.5x)\left(\frac{1}{2}\right)$$

Use the distributive property.

$$2(20) = 2(2.5)(6) + (2.5)x(1)$$

Divide out common factors.

$$40 = 30 + 2.5x$$

Multiply.

$$40 - 30 = 30 + 2.5x - 30$$

Subtract 30 from both sides.

$$10 = 2.5x$$

$$\frac{10}{2.5} = \frac{2.5x}{25}$$

Divide both sides by 2.5.

$$4 = x$$

Joan runs at 4 kilometers per hour and bikes at 2.5(4) = 10 kilometers per hour.

Practice Exercises

1. State the restrictions on the variable in the following rational expressions.

(a) $\dfrac{x-6}{2x-14}$

(b) $\dfrac{4x+1}{x^2-2x-48}$

(c) $\dfrac{5x}{4x^2+9}$

2. Reduce.

(a) $\dfrac{x^2-36}{x+6}$

(b) $\dfrac{2a^2-13a+20}{a^2-16}$

(c) $\dfrac{xb-4x-3b+12}{ab-4a-4b+6}$

(d) $\dfrac{4x^2-25}{5-2x}$

3. Multiply.

(a) $\dfrac{8x^2y}{5xy} \bullet \dfrac{5xy^6}{24x^3y^2}$

(b) $\dfrac{4a-4}{8b^3} \cdot \dfrac{2b}{6a-6}$

(c) $\dfrac{x^2+4x}{x^2+9x+20} \cdot \dfrac{x^2-25}{x^2-3x-10}$

(d)

$$\frac{am+an+bm+bn}{a^2+2ab+b^2} \cdot \frac{a^2-b^2}{ma-mb+an-bn}$$

4. Divide.

(a) $\dfrac{16m^3n}{49m^6} \div \dfrac{24mn^3}{7m^2}$

(b) $\dfrac{x^2-3x}{x^2-5x-14} \div \dfrac{x^2+x-12}{x^2-3x-28}$

(c) $\dfrac{y^3+8}{y^2-4} \div \dfrac{(y^2-2y+4)}{y+2}$

5. Add or subtract as indicated.

(a) $\dfrac{5}{x+1} + \dfrac{x-7}{x+1}$

(b) $\dfrac{y^2+7y-4}{(y+6)\ (y+4)} - \dfrac{2y+2}{(y+6)\ (y+4)}$

(c) $\dfrac{3x-1}{4x+5} - \dfrac{8x+3}{4x+5}$

6. Add or subtract as indicated.

(a) $\dfrac{5}{6a} - \dfrac{7}{12a}$

(b) $\dfrac{x+35}{x^2-2x-15} - \dfrac{24}{x^2-9}$

(c) $6 + \dfrac{4}{2x+5}$

(d) $\dfrac{x-2}{25x^2-10x+1} + \dfrac{1}{5x^2-11x+2}$

(e) $\dfrac{3x^2+7x-40}{x-5} + \dfrac{x^2+10x-5}{5-x}$

7. Simplify.

(a) $\dfrac{\dfrac{6x+9}{4y^2}}{\dfrac{3x+2}{16y^3}}$

(b) $\dfrac{\dfrac{5x-2}{x^2-36}}{\dfrac{25x^2-4}{x+6}}$

(c) $\dfrac{3-\dfrac{5}{x}-\dfrac{2}{x^2}}{3-\dfrac{8}{x}-\dfrac{3}{x^2}}$

(d) $\dfrac{a^{-2}-b^{-2}}{a^{-1}+b^{-1}}$

8. Divide.

(a) $\dfrac{8x^6+24x^5-16x^3}{4x^3}$

(b) $\dfrac{8a^4b^2-9a^3b^3+12a^2}{3a^2b^2}$

9. Divide.

(a) $\dfrac{2x^2+3x-20}{x+5}$

(b) $4x-3$

(c) $\dfrac{a^3-125}{a-5}$

10. Solve each equation.

(a) $\dfrac{x}{2}=\dfrac{3x}{4}+2$

(b) $\dfrac{x+2}{x^2-x}=\dfrac{6}{x^2-1}$

(c) $1-\dfrac{2}{x}=\dfrac{24}{x^2}$

(d) $\dfrac{4}{x^2-9}=\dfrac{5}{x^2+4x+3}+\dfrac{7}{x^2-2x-3}$

11. Solve for the indicated variable.

(a) $V=\dfrac{ah}{3}$ for a

(b) $\dfrac{A}{c}=\dfrac{s}{b}-\dfrac{r}{3}$ for s

(c) $F=\dfrac{km_1m_2}{d^2}$ for k

12. Solve.

(a) If the numerator of the fraction $\frac{4}{7}$ is multiplied by a number and four times that number is added to the denominator, the resulting fraction equals $\frac{20}{27}$

(b) If it takes Joan 4 hours to write a mathematics test while it takes Ron 5 hours, how long will it take them if they write the test together?

(c) A cold water pipe can fill a swimming pool in 10 hours, and the drain takes 15 hours to empty the pool. If the pipe is on and the drain is open, how long will it take to fill the pool?

(d) A professor averages 12 miles per hour biking to the university. On the ride home, he averages 15 miles per hour and bikes an extra 3 miles in the same amount of time. Find the distance from his home to the university.

Answers

1.

(a) $x \neq 7$

(b) $x \neq -6$ and $x \neq 8$

(c) no restrictions

2.

(a) $x - 6$

(b) $\frac{2a-5}{a+4}$

(c) $\frac{x-3}{a-4}$

(d) $-(2x + 5)$ or $-2x - 5$

3.

(a) $\frac{y^4}{x}$

(b) $\frac{1}{6b^2}$

(c) $\frac{x}{x+2}$

(d) 1

4.

(a) $\frac{2}{21m^2n^2}$

(b) $\frac{x}{x+2}$

(c) $\frac{y+2}{y-2}$

5.

(a) $\frac{x-2}{x+1}$

(b) $\frac{y-1}{y+4}$

(c) $\frac{-5x-4}{4x+5}$

6.

(a) $\frac{1}{4a}$

(b) $\frac{x+5}{(x-3)(x-5)}$

(c) $\frac{12x+34}{2x+5}$

(d) $\frac{x^2+x+3}{(5x-1)^2(x-2)}$

(e) $2x + 7$

7.

(a) $\frac{4y(2x+3)}{x+4}$

(b) $\frac{1}{(x-6)(5x+2)}$

(c) $\frac{x-2}{x-3}$

(d) $\frac{b-a}{ab}$

8.

(a) $2x^3 + 6x^2 - 4$

(b) $\frac{8a^2}{3} - 3ab + \frac{4}{b^2}$

9.(a) $2x - 7 + \frac{15}{x+5}$

(b) $4x^2 + 3x + 6 + \frac{6}{4x-3}$

(c) $a^2 + 5a + 25$

10.

(a) -8

(b) 2

(c) $-2, 6$

(d) $-\frac{1}{4}$

(e) -9

(f) -2

11.

(a) $\frac{3V}{h}$

(b) $\frac{3Ab + cbr}{3c}$

(c) $\frac{Fd^2}{m_1 m_2}$

12.

(a) 5

(b) $2\frac{2}{9}$ hours

(c) 30 hours

(d) 12 miles

5

Linear Equations and Inequalities in Two Variables

5.1 GRAPHING USING THE RECTANGULAR COORDINATE SYSTEM

Graphing ordered pairs

When a pair of numbers is written in a specific order in parentheses, such as (2, 5), we call it an **ordered pair**. The numbers written in the ordered pair are called **coordinates**. In (2, 5), 2 is called the ***x*-coordinate** and 5 is called the ***y*-coordinate**.

To graph ordered pairs, we use a rectangular coordinate system made up of two number lines: the horizontal number line is called the x-axis, and the vertical number line is called the y-axis.

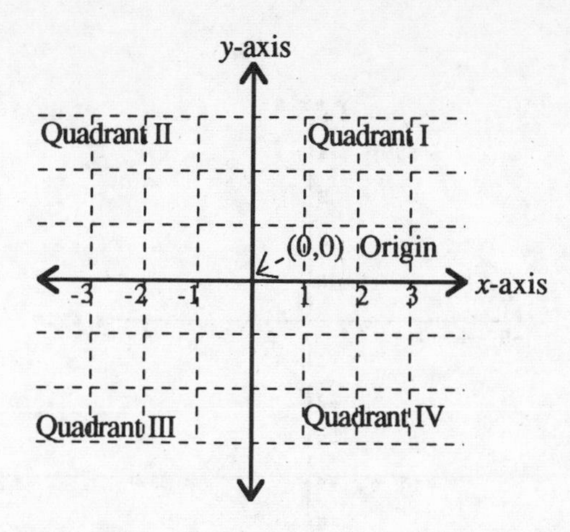

On a number line, positive numbers are to the right of 0, and negative numbers are to the left of 0. Notice that the x-axis is arranged in the same way. On the y-axis, positive numbers are written up the axis, and negative numbers are written down the axis. Where the two axes intersect, both the x- and y-coordinate equal 0. The point (0, 0) is called the **origin**. The four regions formed by the axes are called **quadrants**, and are labeled counter clockwise I, II, III, and IV.

The ordered pair (2, 5) is graphed or plotted by locating the point where $x = 2$ and $y = 5$:

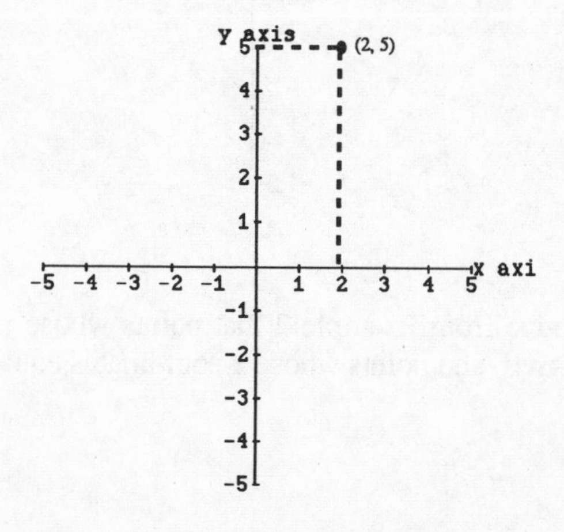

EXAMPLE 1 Plot the points (3, 4), (–3, 4), (–3, –4) and (3, –4) on the same set of axes.

SOLUTION 1

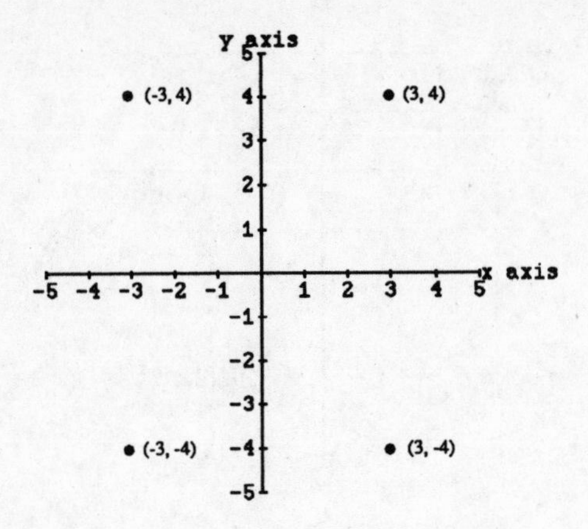

EXAMPLE 2 Plot the points (2, 0), (–4, 0), (0, 1) and (0, –3) on the same set of axes.

SOLUTION 2

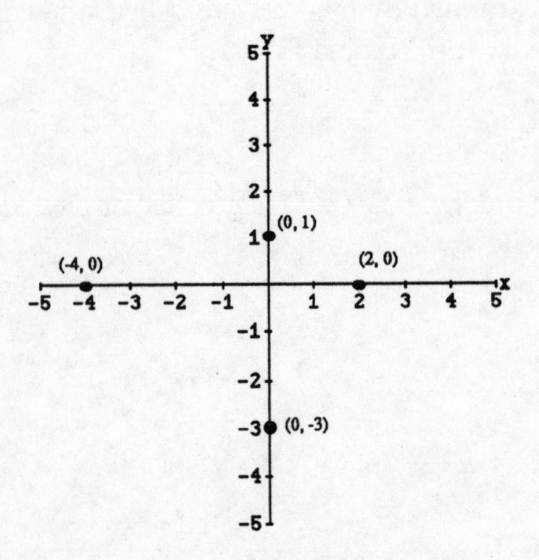

Notice from Example 2 that points whose y-coordinates equal 0 are *on the x-axis*, and points whose x-coordinates equal 0 are *on the y-axis*.

Graphing linear equations using a table of values

A **linear equation in two variables** is an equation that can be written in the form $ax + by = c$, where a, b, and c are real numbers, and a and b are not both 0. A linear equation in two variables has an infinite number of solutions, that is, ordered pairs that satisfy the equation. We use this idea as our first method for graphing linear equations.

To Graph Linear Equations in Two Variables:

1. Find three ordered-pair solutions to the given equation. Completing the following table is often convenient:

x	y
1	
2	
3	

2. Plot the three points.
3. Draw a straight line through these points. If they are not in line, go back and check your arithmetic. Put arrows on both ends of the line to show that the line continues forever in both directions.

Note: Although two points determine a line, we usually plot three as a check on our arithmetic.

EXAMPLE 3 Graph $x - 2y = 5$.

SOLUTION 3

$x - 2y = 5$

Let's complete the table.

x	y
1	
2	
3	

If $x = 1$,

$$1 - 2y = 5$$

	$-2y = 4$	Subtract 1.
	$y = -2$	Divide by -2.

If $x = 2$,

	$2 - 2y = 5$	
	$-2y = 3$	Subtract 2.
	$y = -\frac{3}{2}$	Divide by -2.

If $x = 3$,

	$3 - 2y = 5$	
	$-2y = 2$	Subtract 3.
	$y = -1$	Divide by -2.

The table becomes :

x	y
1	-2
2	-3/2
3	-1

and the graph is $x - 2y = 5$.

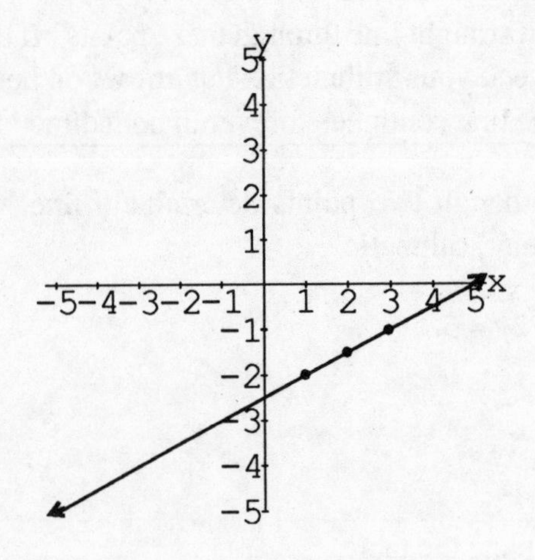

EXAMPLE 4 Graph $3x + y = 4$.

SOLUTION 4

$3x + y = 4$

Let's complete the table.

x	y
1	
2	
3	

If $x = 1$,

$3(1) + y = 4$

$3 + y = 4$ Subtract 3.

$y = 1$

If $x = 2$,

$3(2) + y = 4$

$6 + y = 4$ Subtract 6.

$y = -2$

If $x = 3$,

$3(3) + y = 4$

$9 + y = 4$ Subtract 9.

$y = -5$

The table becomes:

x	y
1	1
2	-2
3	-5

and the graph is $3x + y = 4$.

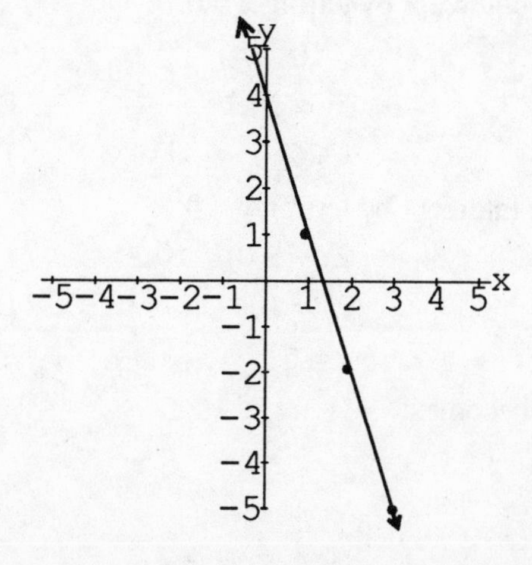

Graphing linear equations using the intercepts

The x-intercept of a graph is the x-coordinate where the graph crosses the x-axis. When plotting points, we noted that this occurs when the y-coordinate is 0. The y-intercept of a graph is the y-coordinate where the graph crosses the y-axis. This occurs when the x-coordinate is 0. A linear equation can be graphed using the intercepts by completing the table:

	x	y	
This is the x-intercept →	0		
		0	← This is the y-intercept

As noted previously, we may use a third point as a check on our arithmetic, often letting $x = 1$.

EXAMPLE 5 Graph $2x + 5y = 10$ using the intercepts.

SOLUTION 5

Let's complete the table.

x	y
0	
	0

Find the y-intercept by letting $x = 0$:

$2(0) + 5y = 10$

$5y = 10$ Divide both sides by 5.

$y = 2$ This is the y-intercept.

Find the x-intercept by letting $y = 0$:

$2x + 5(0) = 10$

$2x = 10$ Divide both sides by 2.

$x = 5$ This is the x-intercept.

The table becomes:

x	y
0	2
5	0

and the graph is $2x + 5y = 10$.

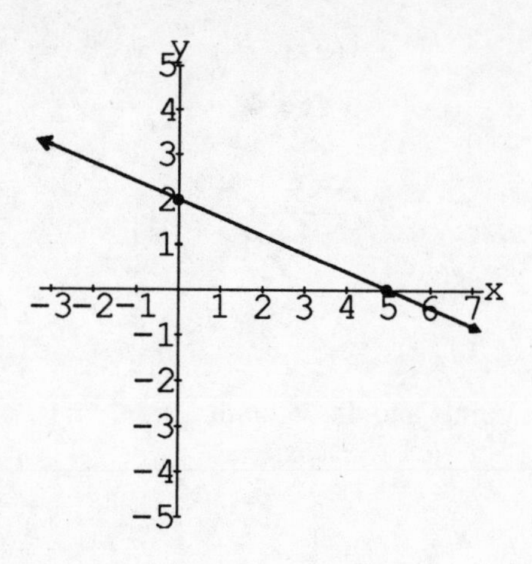

Let's use $x = 1$ to find a check point:

$2(1) + 5y = 10$	Substitute $x = 1$.
$2 + 5y = 10$	Multiply.
$5y = 8$	Subtract 2 from both sides.
$y = \frac{8}{5}$	Divide both sides by 5.

Check the graph to note that $(1, \frac{8}{5})$ is on the line.

EXAMPLE 6 Graph $x - 4y = 0$.

SOLUTION 6

Let's find the x- and y-intercepts.

If $x = 0$,

$0 - 4y = 0$	
$-4y = 0$	Divide both sides by -4.
$y = 0$	This is the y-intercept.

If $y = 0$,

$x - 4(0) = 0$	
$x - 0 = 0$	
$x = 0$	This is the x-intercept.

We have only found one point (0, 0), which is both the x- and y-intercept. To graph the line, find another point by choosing an x-coordinate and finding the y-coordinate.

If $x = 1$,

$$1 - 4y = 0$$

$$-4y = -1 \qquad \text{Subtract 1 from both sides.}$$

$$\frac{-4y}{-4} = \frac{-1}{-4} \qquad \text{Divide both sides by } -4.$$

$$y = \frac{1}{4}$$

Plot the points and draw the line $x - 4y = 0$.

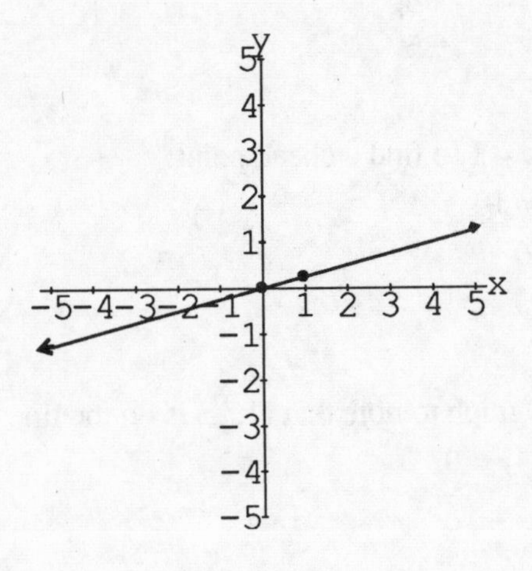

Graphing horizontal and vertical lines

The equation $x = -3$, which can be written $1x + 0y = -3$, has solutions $(-3, 1)$, $(-3, 3)$, $(-3, -2)$, etc., and graphs as a vertical line. The equation $y = 2$ which can be written $0x + y = 2$ has solutions $(3, 2)$, $(2, 2)$, $(-4, 2)$ etc., and graphs as a horizontal line. If you recognize these situations, you do not need a table of values.

> To graph $x = a$, draw a vertical line through the point $(a, 0)$.
> To graph $y = b$, draw a horizontal line through the point $(0, b)$.

EXAMPLE 7 Graph.

a) $x = -3$

b) $y = 2$

SOLUTION 7

a) To graph $x = -3$, locate the point $(-3, 0)$.
Then draw a vertical line through $(-3, 0)$.

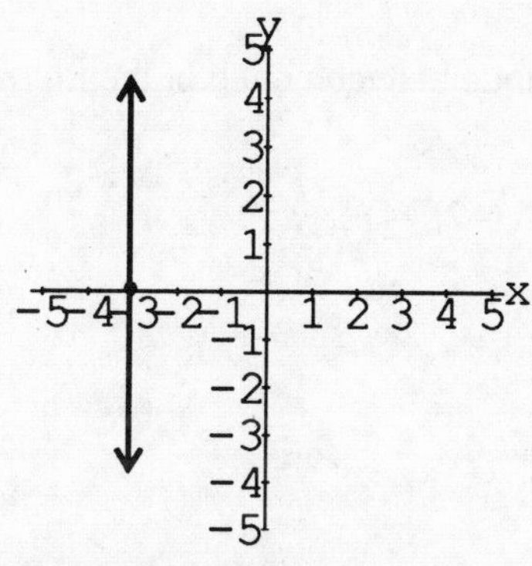

b) To graph $y = 2$, locate the point $(0, 2)$.
Then draw a horizontal line through $(0, 2)$.

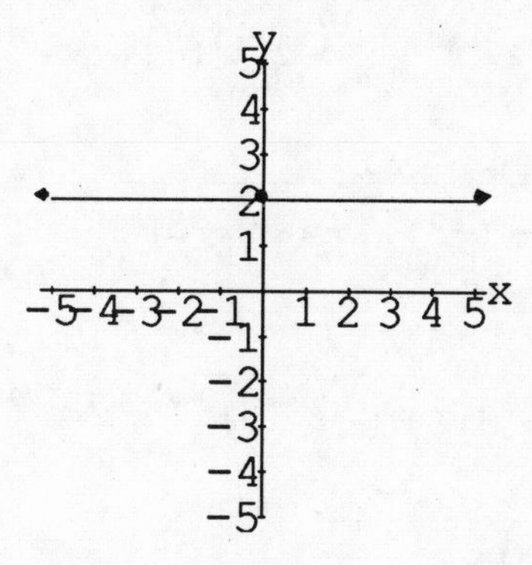

The distance formula

The lines we graphed have infinite length. However, if we consider the part of a line between two points (called a line segment), we can find the length of that line segment using the distance formula.

The Distance Formula

The distance between the points (x_1, y_1) and (x_2, y_2) is

$$d = \sqrt{(x_2 - x_1)^2 + (y_2 - y_1)^2}$$

The small 1 and 2 written slightly below x and y are called subscripts. x_1 is read "x sub 1" and y_2 is read "y sub 2." This notation is helpful when we want to work with several points, but not use specific coordinates.

EXAMPLE 8 Find the distance between each of the following pairs of points.

a) (–3, 5) and (–2, –4)

b) (2, 1) and (5, 5)

SOLUTION 8

a) (–3, 5) and (–2, –4)

$(x_1, y_1) \quad (x_2, y_2)$ — Label the points as (x_1, y_1) and (x_2, y_2).

$d = \sqrt{(x_2 - x_1)^2 + (y_2 - y_1)^2}$ — Write the distance formula.

$= \sqrt{(-2-(-3))^2 + (-4-5)^2}$ — Substitute $x_1 = -3$, $y_1 = 5$, $x_2 = -2$, $y_2 = -4$.

$= \sqrt{(-2+3)^2 + (-4-5)^2}$ — Simplify inside parentheses.

$= \sqrt{(1)^2 + (-9)^2}$ — Simplify under the square root.

$= \sqrt{1+81}$ — Add before trying to take the square root.

$= \sqrt{82}$

If you labeled the points as

$(-3, 5)$ and $(-2, -4)$
(x_1, y_1) (x_2, y_2)

the distance formula would be:

$= \sqrt{(-3-(-2))^2 + (5-(-4))^2}$ Substitute $x_1 = -2, y_1 = -4, x_2 = -3, y_2 = 5$.

$= \sqrt{(-3+2)^2 + (5+4)^2}$ Simplify inside parentheses.

$= \sqrt{(-1)^2 + (9)^2}$ Simplify under the square root.

$= \sqrt{1+81}$ Add before trying to take the square root.

$= \sqrt{82}$

Notice that the distance between $(-3, 5)$ and $(-2, -4)$ is $\sqrt{82}$ regardless of which point you choose as (x_1, y_1) and which you choose as (x_2, y_2).

b) $(2, 1)$ and $(5, 5)$
(x_1, y_1) (x_2, y_2) Label the points as (x_1, y_1) and (x_2, y_2).

$d = \sqrt{(x_2-x_1)^2 + (y_2-y_1)^2}$ Write the distance formula.

$= \sqrt{(5-2)^2 + (5-1)^2}$ Substitute $x_1 = 2, y_1 = 1, x_2 = 5, y_2 = 5$.

$= \sqrt{(3)^2 + (4)^2}$ Simplify inside parentheses.

$= \sqrt{9+16}$ Simplify under the square root.

$= \sqrt{25}$ Simplify the square root.
$= 5$

The midpoint formula

The distance formula enables us to find the length of a line segment. Another useful formula enables us to find the midpoint of a line segment.

The Midpoint Formula
Given (x_1, y_1) and (x_2, y_2), the endpoints of a line segment, the point halfway between the given points is:

$$M = \left(\frac{x_1 + x_2}{2}, \frac{y_1 + y_2}{2}\right)$$

EXAMPLE 9 Find the midpoint of the line segment between the given points.

a) (–3, 5) and (–2, –4)

b) (2, 1) and (5, 5)

SOLUTION 9

a) (–3, 5) and (–2, –4)

(x_1, y_1) (x_2, y_2) — Label the points (x_1, y_1) and (x_2, y_2).

$$M = \left(\frac{x_1 + x_2}{2}, \frac{y_1 + y_2}{2}\right)$$

Write the midpoint formula.

$$= \left(\frac{-3 + (-2)}{2}, \frac{5 + (-4)}{2}\right)$$

Substitute $x_1 = -3$, $y_1 = 5$, $x_2 = -2$, $y_2 = -4$.

$$= \left(\left(-\frac{5}{2}\right), \frac{1}{2}\right)$$

Simplify.

b) (2, 1) and (5, 5)

(x_1, y_1) (x_2, y_2) — Label the points (x_1, y_1) and (x_2, y_2).

$$M = \left(\frac{x_1 + x_2}{2}, \frac{y_1 + y_2}{2}\right)$$

Write the midpoint formula.

$$= \left(\frac{2 + 5}{2}, \frac{1 + 5}{2}\right)$$

Substitute $x_1 = 2$, $y_1 = 1$, $x_2 = 5$, $y_2 = 5$.

$= (\frac{7}{2}, \frac{6}{2})$

$= (\frac{7}{2}, 3)$ Simplify.

Try relabeling the points in Example 9. You will find that the midpoint will still be the same. Compare the answers to Example 8 and Example 9 and note that the solution to a distance is a number, whereas the solution to a midpoint is an ordered pair.

5.2 THE SLOPE OF A LINE

The **slope** of a line is a measure of the steepness of the line. Slope is calculated by the ratio

$$\frac{\text{vertical change}}{\text{horizontal change}}.$$

The slope of a line containing points $P_1(x_1, y_1)$ and $P_2(x_2, y_2)$ is:

$$\text{slope} = m = \frac{rise}{run} = \frac{y_2 - y_1}{x_2 - x_1} = \frac{y_1 - y_2}{x_1 - x_2}$$

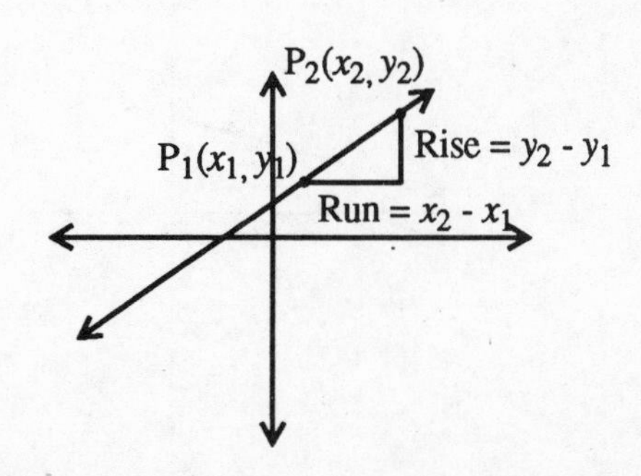

Notice that we describe the steepness as "rise" over "run." Where the slope is negative, we say that the slope "falls" rather than rises.

EXAMPLE 10 Find the slope of each of the following lines.

a) the line through (1, 2) and (5, 3)

b) the line through (–1, 5) and (4, 3)

SOLUTION 10

a) (1, 2) (5, 3)

(x_1, y_1) (x_2, y_2) — Label the points (x_1, y_1) and (x_2, y_2).

$m = \dfrac{y_2 - y_1}{x_2 - x_1}$ — Write the formula for slope.

$= \dfrac{3-2}{5-1}$ — Substitute $x_1 = 1$, $y_1 = 2$, $x_2 = 5$, $y_2 = 3$.

$= \dfrac{1}{4}$ — Simplify the numerator and denominator.

This line has positive slope and *rises* from left to right.

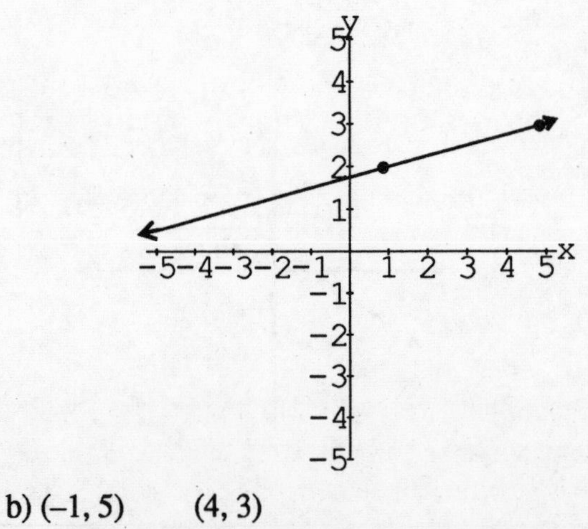

b) (–1, 5) (4, 3)

(x_1, y_1) (x_2, y_2) — Label the points (x_1, y_1)

and (x_2, y_2).

$m = \dfrac{y_2 - y_1}{x_2 - x_1}$ Write the formula for slope.

$= \dfrac{3-5}{4-(-1)}$ Substitute $x_1 = -1$, $y_1 = 5$, $x_2 = 4$, $y_2 = 3$.

$= -\dfrac{2}{5}$ Simplify the numerator and denominator.

This line has negative slope and *falls* from left to right.

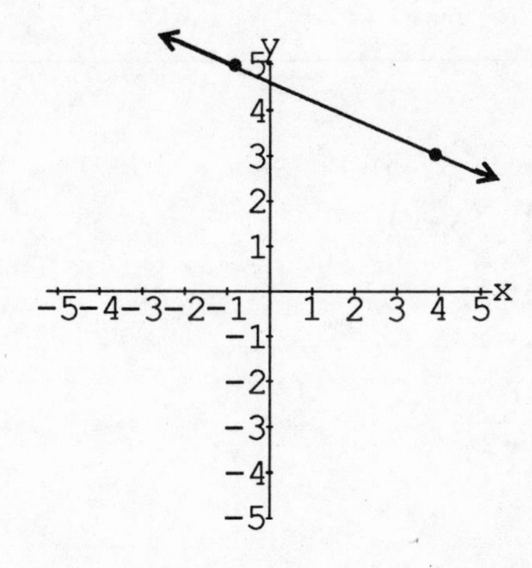

In example 10, you could have labeled either point as (x_1, y_1), the slope will stay the same.

Warnings:
1. Always find the change in ***y*** in the numerator of the formula for slope.
2. Always label the points with (x_n, y_n) not (x_1, x_2) or (y_1, y_2).

Slope of horizontal and vertical lines

Slope measures the steepness of a line. Horizontal lines do not rise, and have a slope of 0. Vertical lines have no run, and the slope formula becomes $\dfrac{run}{0}$, which is undefined. The next example demonstrates two points to remember.

The slope of any horizontal line is 0. The slope of any vertical line is undefined.

EXAMPLE 11 Find the slope of each of the following lines.

a) The line through (2, –3) and (–1, –3).

b) The line through (1, –1) and (1, 4).

SOLUTION 11

a) (2, –3) and (–1, –3)

(x_1, y_1) (x_2, y_2) — Label the points (x_1, y_1) and (x_2, y_2).

$m = \dfrac{y_2 - y_1}{x_2 - x_1}$ — Write the formula for slope.

$= \dfrac{-3-(-3)}{-1-2}$ — Substitute $x_1 = 2$, $y_1 = -3$, $x_2 = -1$, $y_2 = -3$.

$= \dfrac{0}{-3} = 0$ — Simplify the fraction.

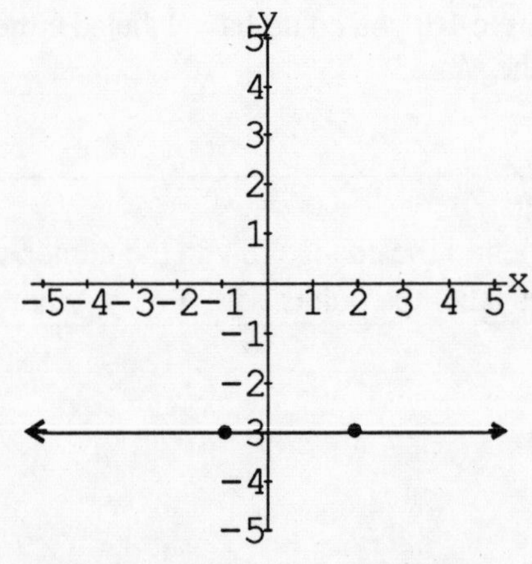

b) (1, –1) and (1, 4)

(x_1, y_1) (x_2, y_2) — Label the points (x_1, y_1) and (x_2, y_2).

$= \dfrac{4-(-1)}{1-1}$ — Substitute $x_1 = 2$, $y_1 = -3$, $x_2 = -1$, $y_2 = -3$.

$= \frac{5}{0} =$ undefined

Simplify the fraction. Vertical lines have undefined slope.

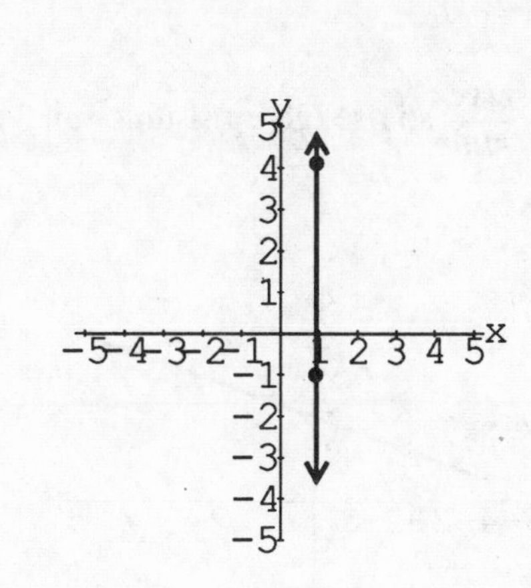

Graphing using the slope and a point

In section 5.1 we graphed lines after finding at least two points on the line. We can also graph a line if we know one point on the line and the slope of the line.

To Graph the Line through the Point (x_1, y_1) with Slope $m = \frac{a}{b}$:

1. Put a point on (x_1, y_1) and use this as your starting point.
2. Use $m = \frac{rise}{run} = \frac{a}{b}$ and rise (or go up) *a* units from your starting point.

   Then move to the right *b* units and put a second point there.
3. Draw a line through the two points.

Note: If the given slope is an integer, write it as a fraction using 1 for the denominator. If the given slope is negative, carry the negative in the numerator and fall (go down) *a* units instead of rising *a* units.

EXAMPLE 12 Graph the line through (–2, 1) with slope $\frac{1}{3}$.

SOLUTION 12

Start at (–2, 1).

$m = \frac{1}{3} = \frac{rise}{run}$ so rise (go up) 1 unit, run 3 units to the right.

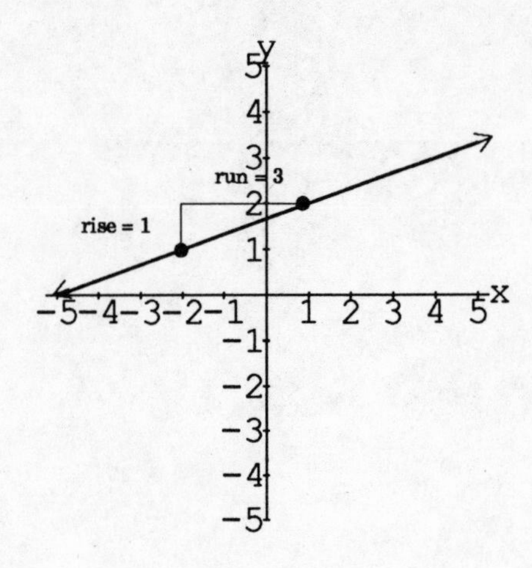

EXAMPLE 13 Graph the line through (0, 5) with slope –4.

SOLUTION 13

Start at (0, 5).

$m = -4 = \frac{-4}{1}$ Write –4 as $\frac{-4}{1}$.

$m = \frac{-4}{1} = \frac{rise}{run}$ so fall (go down) 4 units and run to the right 1 unit.

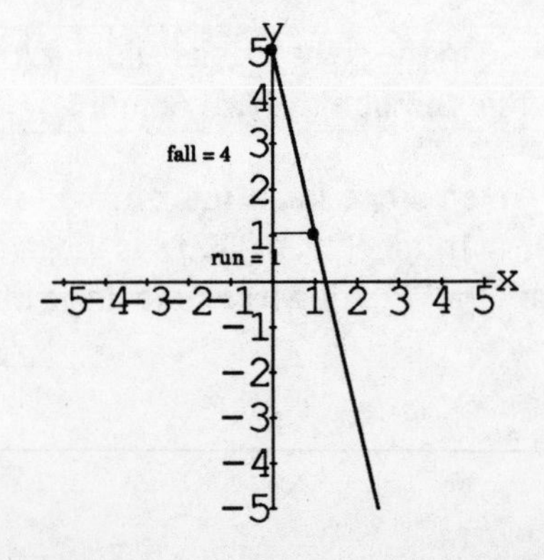

Slopes of parallel and perpendicular lines

Perhaps you recall studying parallel and perpendicular lines in geometry. Parallel lines are lines in geometry. Parallel lines are lines in the same plane that do not intersect. Perpendicular lines are lines in the same plane that intersect at right angles. The slopes of parallel lines and perpendicular lines have special relationships.

> Parallel lines have equal slopes.
> Perpendicular lines have slopes that are negative reciprocals of each other (a and $-\frac{1}{a}$, for $a \neq 0$).

EXAMPLE 14 Determine whether the line L_1 through (4, 3) and (–8, –6) is parallel or perpendicular to the line L_2 through (2, 1) and (–6, –5).

SOLUTION 14

Find the slope of L_1:

(4, 3) (–8, –6)

(x_1, y_1) (x_2, y_2) Label the points (x_1, y_1) and (x_2, y_2).

$m = \dfrac{y_2 - y_1}{x_2 - x_1}$ Write the formula for slope.

$= \dfrac{-6-3}{-8-4}$ Substitute $x_1 = 4$, $y_1 = 3$, $x_2 = -8$, $y_2 = -6$.

$= \dfrac{-9}{-12}$ Simplify the fraction.

$= \dfrac{3}{4}$ Slope of L_1.

Find the slope of L_2:

(2, 1) (–6, –5)

(x_1, y_1) (x_2, y_2) Label the points (x_1, y_1) and (x_2, y_2).

$m = \dfrac{y_2 - y_1}{x_2 - x_1}$ Write the formula for slope.

$= \dfrac{-5-1}{-6-2}$ Substitute $x_1 = 2, y_1 = 1,$ $x_2 = -6, y_2 = -5.$

$= \dfrac{-6}{-8}$ Simplify the fraction.

$= \dfrac{3}{4}$ Slope of L_2.

Since the slope of L_1 = slope of $L_2 = \frac{3}{4}$, L_1 and L_2 are parallel lines.

EXAMPLE 15 Find a so that the slope of L_1 through $(a, -7)$ and $(-1, 3)$ is perpendicular to L_2 with slope $\frac{2}{5}$.

SOLUTION 15

Find the slope of L_1: $(a, -7)$ $(-1, 3)$

(x_1, y_1) (x_2, y_2) Label the points (x_1, y_1) and (x_2, y_2).

$m = \dfrac{y_2 - y_1}{x_2 - x_1}$ Write the formula for slope.

$= \dfrac{3-(-7)}{-1-a}$ Substitute $x_1 = a, y_1 = -7,$ $x_2 = -1, y_2 = 3.$

$= \dfrac{10}{-1-a}$ Slope of L_1.

L_1 will be perpendicular to L_2 if its slope is the negative reciprocal of $\frac{2}{5}$ or $-\frac{5}{2}$.

$\dfrac{10}{-1-a} = -\dfrac{5}{2}$ Set slope of L_1 equal to $-\frac{5}{2}$.

$2(-1-a)\left(\dfrac{10}{-1-a}\right) = 2(-1-a)\left(-\dfrac{5}{2}\right)$ Multiply both sides by the LCD.

$2(10) = (-1-a)(-5)$ Divide out common factors.

$20 = 5 + 5a$ Multiply.

$20 - 5 = 5 + 5a - 5$ Subtract 5 from both sides.

$15 = 5a$ Simplify both sides.

$3 = a$ Divide both sides by 5.

5.3 FINDING THE EQUATION OF A LINE

Writing equations of lines

In this section we will find equations of lines, given various pieces of information. The table below will help you decide when to use each form of a linear equation. You must memorize these forms!

Given This Data	Use This Form	Name of the Form
Slope m and y-intercept b	$y = mx + b$	Slope-intercept
Point (x_1, y_1) and slope m	$y - y_1 = m(x - x_1)$	Point-slope
Two points (x_1, y_1) and (x_2, y_2)	First find $m = \frac{y_2 - y_1}{x_2 - x_1}$ then use: $y - y_1 = m(x - x_1)$	Slope of a line Point-slope
Horizontal line through (x_1, y_1)	$y = y_1$	
Vertical line through (x_1, y_1)	$x = x_1$	

EXAMPLE 16 Write an equation of the line using the given information.

a) $m = 2$, y-intercept $= 1$

b) $m = \frac{3}{2}$, containing the point $(-1, -2)$

c) containing the points $(-1, 2)$ and $(2, 1)$

d) a horizontal line passing through $(3, 4)$

e) a vertical line passing through $(3, 4)$

SOLUTION 16

a) $m = 2, b = 1$ — Given slope and y-intercept.

$y = mx + b$ — Slope-intercept form.

$y = 2x + 1$ — Substitute $m = 2$ and $b = 1$ into the slope-intercept form of a line.

b) $m = \frac{3}{2}, (-1, -2)$ — Given slope and a point.

$y - y_1 = m(x - x_1)$ — Point-slope form.

$y - (-2) = \frac{3}{2}(x - (-1))$ — Substitute $m = \frac{3}{2}$, $x_1 = -1$, $y_1 = -2$.

$y - 2 = \frac{3}{2}(x + 1)$ — Equation of the line.

c) (x_1, y_1) (x_2, y_2) — Given two points.

$(-1, 2)$ and $(2, 1)$

$m = \frac{y_2 - y_1}{x_2 - x_1} = \frac{1 - 2}{2 - (-1)}$ — First find the slope.

$m = -\frac{1}{3}$

$y - y_1 = m(x - x_1)$ — Now use the point-slope form of a line.

$y - 2 = -\frac{1}{3}(x - (-1))$ — Substitute $y_1 = 2$, $m = -\frac{1}{3}$, and $x_1 = -1$.

$y - 2 = -\frac{1}{3}(x + 1)$ — Equation of the line.

d) Horizontal line through (3, 4)

$y = 4$ — Equation of line.

e) Vertical line through (3, 4)

$x = 3$ — Equation of line.

Forms of linear equations

All the answers given in Example 16 are linear equations in two variables. You may need to write your answers in slope-intercept or standard form. After you have found the equation, use the Addition and Multiplication Properties of Equality to rewrite the equation in the desired form

Forms of Linear Equations	
$ax + by = c$	Standard form
$y = mx + b$	Slope-intercept form

EXAMPLE 17 Rewrite each equation in the required form.

a) $y = 2x + 1$ in standard form

b) $y + 2 = \frac{3}{2}(x + 1)$ in slope-intercept form

SOLUTION 17

a) $y = 2x + 1$ — We need $ax + by = c$.

$y - 2x = 2x + 1 - 2x$ — Subtract $2x$ from both sides. Use the commutative property to rewrite $y - 2x$ as $-2x + y$.

$2x - y = -1$ — Multiply both sides by -1.

b) $y + 2 = \frac{3}{2}(x + 1)$ — We need $y = mx + b$.

$y + 2 = \frac{3}{2}x + \frac{3}{2}$ — Distribute $\frac{3}{2}$.

$y + 2 - 2 = \frac{3}{2}x + \frac{3}{2} - 2$ — Subtract 2 from both sides.

$y = \frac{3}{2}x - \frac{1}{2}$ — Simplify both sides.

Using slope-intercept form for graphing

After an equation has been written in slope-intercept form, we can easily graph the equation. Use the y-intercept (b) as the starting point, and then use $m = \frac{rise}{run}$ to find a second point on the line.

EXAMPLE 18 Graph each equation using the slope and y-intercept.

a) $2x + 5y = 10$

b) $3x - y = 4$

SOLUTION 18

a) $2x + 5y = 10$ — Write the equation in $y = mx + b$ form.

$2x + 5y - 2x = -2x + 10$ — Subtract $2x$ from both sides.

$5y = -2x + 10$ — Combine similar terms.

$\frac{1}{5}(5y) = \frac{1}{5}(-2x + 10)$ — Multiply both sides by $\frac{1}{5}$.

$y = -\frac{2}{5}x + 2$ — Use the distributive property.

$m = -\frac{2}{5}, b = 2$ — Identify m and b.

Start at (0, 2). $m = -\frac{2}{5} = \frac{rise}{run}$ so fall 2 units and run 5 units.

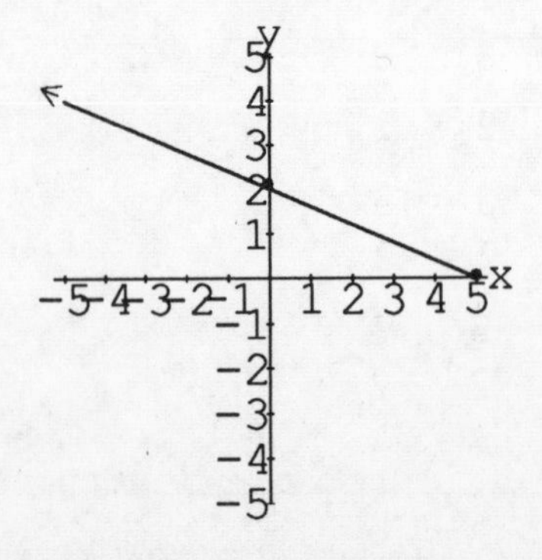

b) $3x - y = 4$ — Write the equation in $y = mx + b$ form.

$3x - y - 3x = -3x + 4$	Subtract $3x$ from both sides.
$-y = -3x + 4$	Combine similar terms.
$y = 3x - 4$	Multiply both sides by -1.
$m = 3 = \frac{3}{1}, b = -4$	Identify m and b.

Start at $(0, -4)$. $m = \frac{3}{1} = \frac{rise}{run}$, so rise 3 units and run 1 unit.

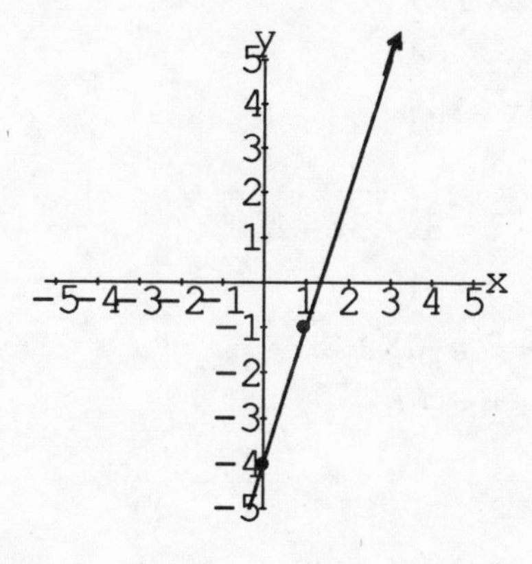

Using slope-intercept for finding parallel and perpendicular lines

We can use the slope-intercept form to help find the equation of a line parallel or perpendicular to a given line.

EXAMPLE 19 Find the equation of the line parallel to $3x + y = -4$ and through $(-6, 2)$. Write the equation in standard form.

SOLUTION 19

Any line parallel to $3x + y = -4$ has the same slope. To find the slope of $3x + y = -4$, rewrite the equation in slope-intercept form:

$3x + y - 3x = -3x - 4$	Subtract $3x$ from both sides.
$y = -3x - 4$	$y = mx + b$ form.

The parallel line must have $m = -3$, and pass through $(-6, 2)$. Sincewehave a point and the slope, use the point-slope formula:

$y - y_1 = m(x - x_1)$	Write the formula.
$y - 2 = -3(x - (-6))$	Substitute $x_1 = -6$, $y_1 = 2$, $m = -3$.
$y - 2 = -3(x + 6)$	Simplify inside the parentheses.
$y - 2 = -3x - 18$	Use the distributive property.
$3x + y - 2 = 3x - 3x - 18$	Add $3x$ to both sides.
$3x + y - 2 = -18$	Combine similar terms.
$3x + y - 2 + 2 = -18 + 2$	Add 2 to both sides.
$3x + y = -16$	The equation is in standard form.

EXAMPLE 20 Find the equation of the line perpendicular to $3x - 5y = 10$ and through $(-1, -4)$. Write the equation in slope-intercept form.

SOLUTION 20

First find the slope of $3x - 5y = 10$:

$3x - 5y - 3x = -3x + 10$	Subtract $3x$ from both sides.
$-5y = -3x + 10$	Combine similar terms.
$-\frac{1}{5}(-5y) = -\frac{1}{5}(-3x + 10)$	Multiply both sides by $-\frac{1}{5}$.
$y = \frac{3}{5}x - 2$	Multiply.

The slope of this line is $m = \frac{3}{5}$. Any line perpendicular to this line must haveslope $m = -\frac{5}{3}$. Since we have a point and the slope, use the point-slope formula:

$y - y_1 = m(x - x_1)$	Write the formula.
$y - (-4) = -\frac{5}{3}(x - (-1))$	Substitute $x_1 = -1$, $y_1 = -4$, $m = -\frac{5}{3}$.
$y + 4 = -\frac{5}{3}(x + 1)$	Simplify each side of the equation.
$y + 4 = -\frac{5}{3}x - \frac{5}{3}$	Use the distributive property.
$y + 4 - 4 = -\frac{5}{3}x - \frac{5}{3} - 4$	Subtract 4 from both sides.
$y = -\frac{5}{3}x - \frac{5}{3} - \frac{12}{3}$	Combine similar terms.
$y = -\frac{5}{3}x - \frac{17}{3}$	The equation is in slope-intercept form.

5.4 GRAPHING LINEAR INEQUALITIES AND SYSTEM OF LINEAR INEQUALITIES

Graphing linear inequalities in two variables involves shading a part of the plane bounded by a line. The line may be solid or dotted, and the shading may be above or below the boundary, depending on the inequality symbol.

Graphing linear inequalities in two variables

To Graph a Linear Inequality in Two Variables:

1. Rewrite the inequality in the form $y\square\ mx + b$, where the box contains $<, >, \leq,$ or $\geq$.
2. Graph a solid line at $y = mx + b$ if the box contains $\leq$ or $\geq$.
   Graph a dotted line at $y = mx + b$ if the box contains $<$ or $>$,
3. Shade above the line if the box contains $>$ or $\geq$.
   Shade below the line if the box contains $<$ or $\leq$.

If you prefer to use a test point to decide where to shade, pick a point on either side of the line, substitute the values in original equation. If the resulting statement is true, shade the side of the line containing the test point. If the resulting statement is false, shade the opposite side of the line.

EXAMPLE 21 Graph each inequality.

a) $3x + 2y < 8$

b) $4x - 3y \leq 6$

SOLUTION 21

a) $3x + 2y < 8$

$3x + 2y - 3x < -3x + 8$ Subtract $3x$ from both sides.

$2y < -3x + 8$ Combine similar terms.

$\frac{1}{2}(2y) < \frac{1}{2}(-3x + 8)$ Multiply both sides by $\frac{1}{2}$.

$y < -\frac{3}{2}x + 4$

The line will be dotted because the inequality symbol is $<$. The shading will be below the line because the inequality symbol is $<$. Graph a dotted line with $m = -\frac{3}{2}$ and $b = 4$.

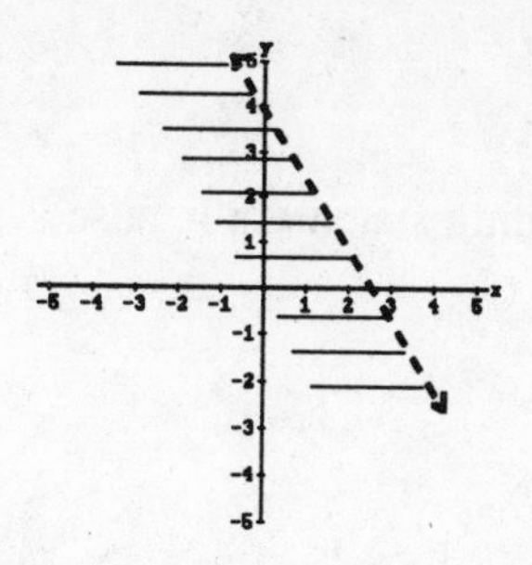

Test point: Let's check (0, 0) in the original inequality.

$3x + 2y < 8$	Original inequality.
$3(0) + 2(0) < 8$	Substitute $x = 0, y = 0$.
$0 < 8$	True statement.

Notice that the side of the line containing (0, 0) has been shaded.

b) $4x - 3y \le 6$

$4x - 3y - 4x \le -4x + 6$	Subtract $4x$ from both sides.
$-3y \le -4x + 6$	Combine similar terms.
$-\frac{1}{3}(-3y) \ge -\frac{1}{3}(-4x + 6)$	Multiply both sides by $-\frac{1}{3}$. **Note**: Reverse the inequality symbol when multiplying by a negative.
$y \ge \frac{4}{3}x - 2$	

The line will be solid because the inequality is $\ge$. The shading will be above the line because the inequality symbol is $\ge$. Graph a solid line with $m = \frac{4}{3}$ and $b = -2$.

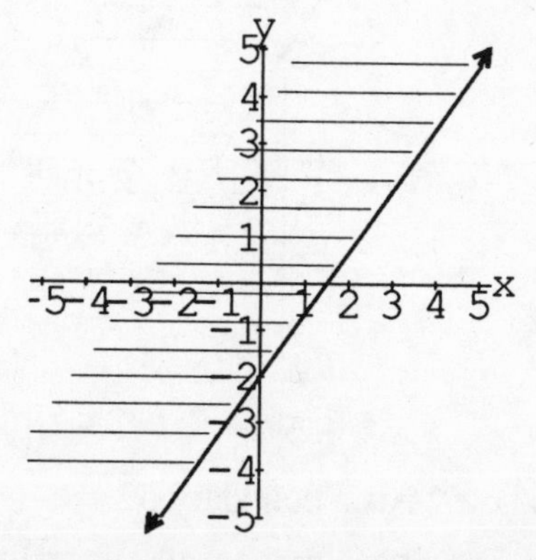

Test point: Let's check (3, 0) in the *original* inequality:

$4x - 3y \leq 6$	Original inequality.
$4(3) - 3(0) \leq 6$	Substitute $x = 3, y = 0$.
$12 \leq 6$	False statement.

Since the resulting statement is false, shade the opposite side of the line.

Graphing systems of inequalities

Our work in Chapter 2 included graphing intersections and unions of inequalities. The word "and" indicates that an intersection is required and the word "or" indicates that a union is required.

EXAMPLE 22 Graph the intersection of $3x + 2y \geq 4$ and $x \geq 2$.

SOLUTION 22

Graph each inequality separately. If you have colored pencils, it may help to graph each inequality in a different color.

$3x + 2y \geq 4$	Rewrite in $y\square mx + b$ form.
$2y \geq -3x + 4$	Subtract $3x$ from both sides.
$y \geq -\frac{3}{2}x + 2$	Multiply both sides by $\frac{1}{2}$.

Graph a solid line with $m = -\frac{3}{2}$, $b = 2$ and shade above the line:

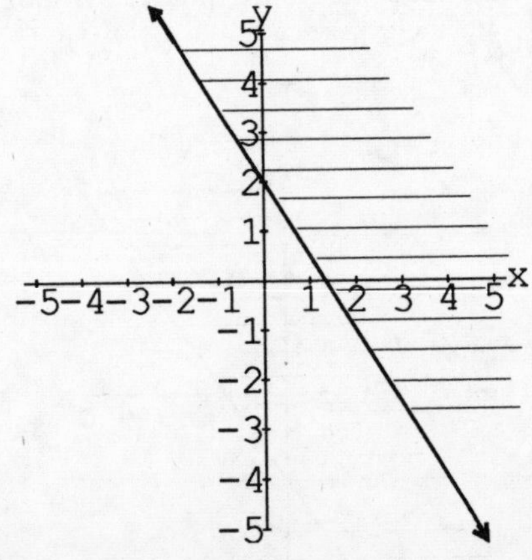

The graph of $x \geq 2$ has a boundary of $x = -1$ which is a vertical line through the point (–1, 0). Shade to the right of the line to represent the points where $x > 2$.

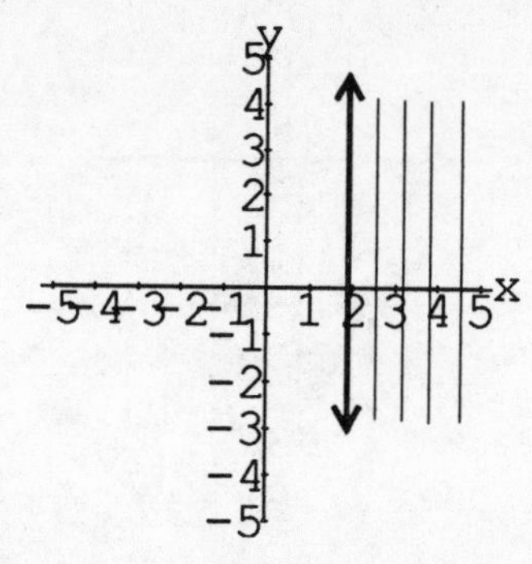

The intersection occurs where the shading overlaps:

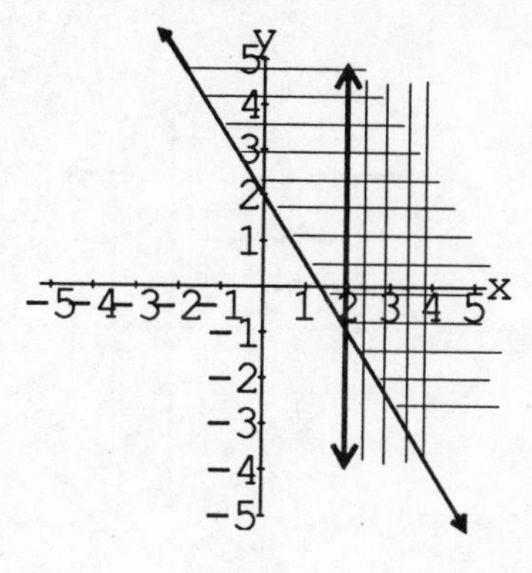

EXAMPLE 23 Graph the union of $x \leq 2$ and $y \leq 3$.

SOLUTION 23

Graph each inequality separately. $x \leq 2$ has a boundary of $x = 2$ which is a vertical line through (2, 0). Shade to the left of the line to represent the points where $x < 2$:

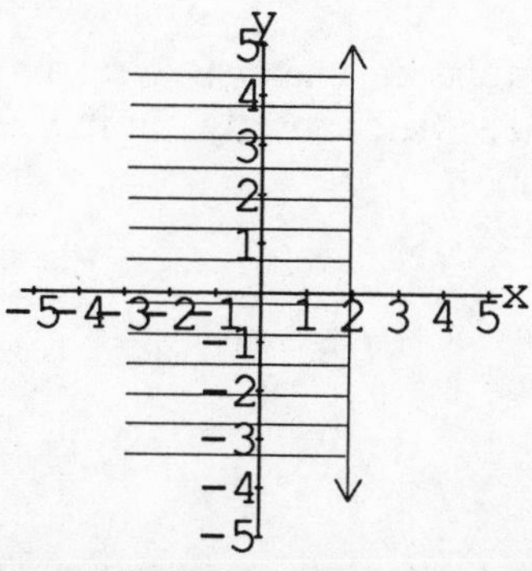

$y \leq 3$ has a boundary of $y = 3$ which is a horizontal line through (0, 3). Shade below the line to represent the points where $y < 3$:

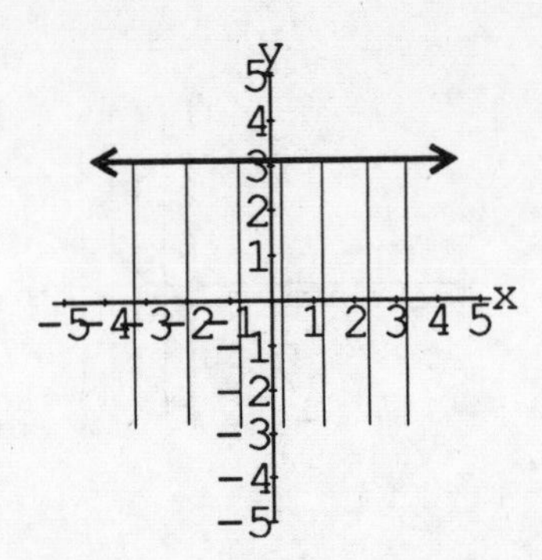

The union of these two graphs contains *all* the points previously shaded:

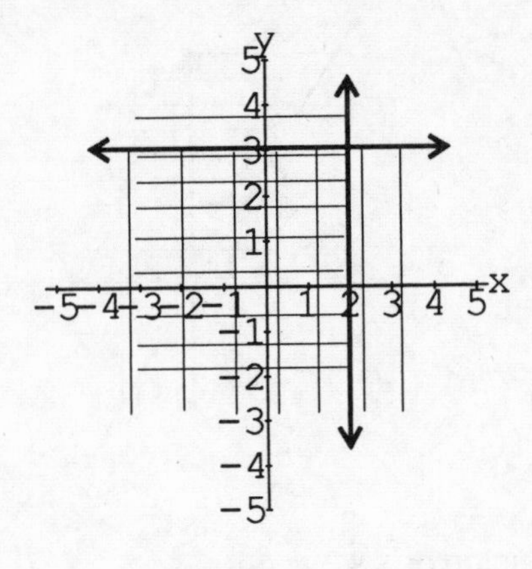

5.5 VARIATION

Many formulas in chemistry and physics involve variables that vary **directly, indirectly**, or **jointly**. In this section we will discuss these three types of variation.

Basic formulas

Each type of variation involves a number called the **constant of variation** and **constant of proportionality**, usually represented by the letter k. The following table summarizes the basic formulas for variation.

Type of Variation	Equation	English Phrase
Direct variation	$y = kx$	y varies directly as x. y is proportional to x.
Indirect variation	$y = \frac{k}{x}$	y varies indirectly as x. or y varies inversely as x.
Joint variation	$y = kxz$	y varies jointly as x and z.

Variation problems generally require a two-step procedure. First k, the constant of variation, is found, and then that value of k is used to find a missing variable.

Direct variation

In a direct variation, as one variable increases, so does the other. As one variable decreases, so does the other.

EXAMPLE 24 y varies directly as x. If x is 30 when y is 180, find y when x is 50.

SOLUTION 24

$y = kx$ — Write the equation for direct variation.

Step 1.

Find k.

$180 = k(30)$ — Substitute $y = 180$ and $x = 30$.

$\frac{180}{30} = \frac{k(30)}{30}$ — Divide both sides by 30.

$6 = k$

Step 2.

Find y when x is 50.

$y = 6x$ — Write the equation with $k = 6$.

$y = 6(50)$ — Substitute $x = 50$.

$y = 300$

Thus y is 300 when x is 50.

EXAMPLE 25 Hooke's Law states that the force needed to stretch a spring is directly proportional to the distance the spring is stretched. If a force of 3 pounds stretches a certain spring 2 centimeters, find the force needed to stretch the spring 16 centimeters.

SOLUTION 25

Let

F = the force and d = the distance the spring is stretched

then

$F = kd$ — Write the equation for direct variation.

Step 1. Find k.

$3 = k(2)$ — Substitute $F = 3$ and $d = 2$.

$\frac{3}{2} = \frac{k(2)}{2}$ — Divide both sides by 2.

$6 = k$

Step 2. Find F when d is 16.

$F = \frac{3}{2}d$ — Write the equation with $k = \frac{3}{2}$.

$F = \frac{3}{2}(16)$ — Substitute $d = 16$.

$F = 24.$

Thus a force of 24 pounds is needed to stretch the spring 16 centimeters.

Indirect variation

In an indirect variation, as one variable increases, the other variable decreases.

EXAMPLE 26 y varies inversely as x. If $y = 25$ when $x = 4$, find y when $x = 20$.

SOLUTION 26

$$y = \frac{k}{x}$$ Write the equation for indirect variation.

Step 1. Find k.

$$25 = \frac{k}{4}$$ Substitute $y = 25$ and $x = 4$.

$$4 \cdot 25 = 4 \cdot \frac{k}{4}$$ Multiply both sides by 4.

$$100 = k$$

Step 2. Find y when x is 20.

$$y = \frac{100}{x}$$ Write the equation with $k = 100$.

$$y = \frac{100}{20}$$ Substitute $x = 20$.

$$y = 5$$

Thus y is 5 when x is 20.

EXAMPLE 27 Boyle's Law states that the pressure of a compressed gas is inversely proportional to the volume of the gas. If there is a pressure of 36 pounds per square inch when the volume of gas is 50 cubic inches, find the pressure when the gas has a volume of 100 cubic inches.

SOLUTION 27

Let P = pressure of the gas
and V = volume of the gas

$$P = \frac{k}{V}$$ Write the equation for indirect variation.

Step 1. Find k.

$36 = \frac{k}{50}$ — Substitute $P = 36$ and $V = 50$.

$50 \cdot 36 = 50 \cdot \frac{k}{50}$ — Multiply both sides by 50.

$1800 = k$

Step 2. Find P when V is 100.

$P = \frac{1800}{k}$ — Write the equation with $k = 1800$.

$P = \frac{1800}{100}$ — Substitute $V = 100$.

$P = 18$

Thus the pressure is 18 pounds per square inch when the volume is 100 cubic inches. Note that as the volume *increased*, the pressure *decreased.*

Joint variation

If the value of y changes as the value of x and z change, we say y varies jointly with x and z.

EXAMPLE 28 y varies jointly as x^2 and z. If y is 24 when x is 2 and z is 3, find y when x is 6 and z is 4.

SOLUTION 28

$y = kx^2z$ — Write the equation for joint variation.

Step 1. Find k.

$24 = k(2)^2(3)$ — Substitute $y = 24$, $x = 2$, and $z = 3$.

$24 = k(12)$ — Simplify the right side.

$\frac{24}{5} = \frac{k(12)}{12}$ — Divide both sides by 12.

$2 = k$

Step 2. Find y when x is 6 and y is 4.

$y = 2x^2z$ — Write the equation with

	$k = 2.$
$y = 2(6)^2(4)$	Substitute $x = 6$ and $z = 4$.
$y = 2(36)(4)$	Simplify the right side.
$y = 288$	

Thus y is 288 when x is 6 and y is 4.

EXAMPLE 29 The volume of a right circular cylinder varies jointly as the height and the square of the radius. If the volume is 45π cubic inches when the radius is 3 inches and the height is 5 inches, find the volume when the height is 2.5 inches and the radius is 10 inches.

SOLUTION 29

Let

V = volume

h = height

r = radius

Then

$V = khr^2$	Write the equation for joint variation.

Step 1. Find k.

$45\pi = k(5)(3^2)$	Substitute $V = 45\pi$, $h = 5$ and $r = 3$.
$45\pi = k(45)$	Simplify the right side.
$\frac{45\Pi}{45} = \frac{k(45)}{45}$	Divide both sides by 45.
$\pi = k$	

Step 2. Find V when $h = 2.5$ and $r = 10$.

$V = \pi hr^2$	Write the equation with
$k = \pi.$	
$V = \pi(2.5)(10^2)$	Substitute $h = 2.5$ and $r = 10$.
$V = \pi(250)$	Simplify the right side.
$V = 250\pi$	

Thus the volume is 250π cubic inches when the height is 2.5 inches and the radius is 10 inches.

Practice Exercises

1. Plot the points (3, 2), (–3, 2), (–3, –2), and (3, –2) on the same set of axes.
2. Plot the points (3, 0), (0, 4), (–1, 0), and (0, –2) on the same set of axes.

3. Graph.

(a) $2x - 3y = 6$

(b) $2x + y = 4$

(c) $2x - 5y = 0$

4. Find the x- and y-intercepts.

(a) $2x + 4y = 8$

(b) $3x - y = 6$

5. Graph.

(a) $x = 4$

(b) $y = -1$

6. Find the distance between each pair of points.

(a) (6, –1) and (–2, 4)

(b) (–4, 7) and (1, –3)

7. Find the midpoint of the line segment between each pair of points.

(a) (6, –1) and (–2, 4)

(b) (–4, 7) and (–1, –3)

8. Find the slope of each of the following lines.

(a) (2, 7) and (–4, –2)

(b) (–1, –3) and (5, –2)

(c) (4, 2) and (–6, 2)

(d) (–5, 1) and (–5, –3)

9. Determine if the lines L_1 and L_2 passing through the given pairs of points are parallel, perpendicular, or neither.

(a) L_1: (–2, –3), (1, 3)
L_2: (4, 5), (4, –11)

(b) L_1: (–2, –4), (2, –2)
L_2: (3, –6), (5, –10)

(c) L_1: (–1, –2), (3, 5)
L_2: (–2, 3), (5, 7)

10. Write an equation of the line using the given information. Write each answer in standard form.

(a) $m = -3$, y-intercept = 2

(b) $m = \frac{1}{2}$, containing the point (–2, 3)

(c) containing the points (–3, –2) and (4, –6)

11. Write an equation of the line using the given information. Write each answer in slope-intercept form.

(a) $m = -\frac{2}{3}$, y-intercept = 6

(b) $m = 4$, containing the point (–1, 5)

12. Write an equation of the line using the given information.

(a) a horizontal line passing through (–1, 2)

(b) a vertical line passing through (–4, 6)

13. Write each equation in slope-intercept form and then graph.

(a) $6x + 2y = 4$

(b) $x - 3y = 9$

Solutions

1.

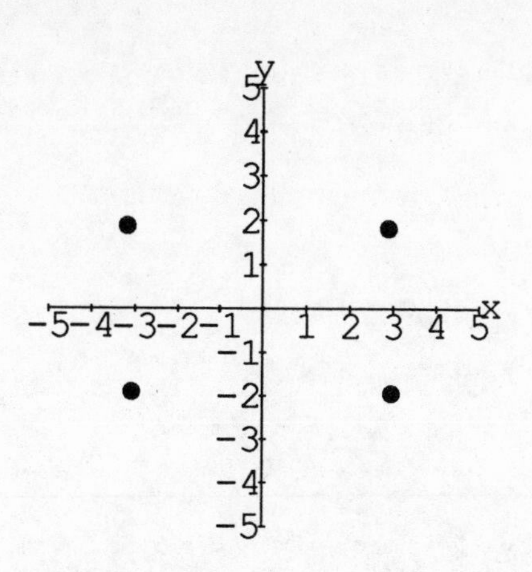

2.

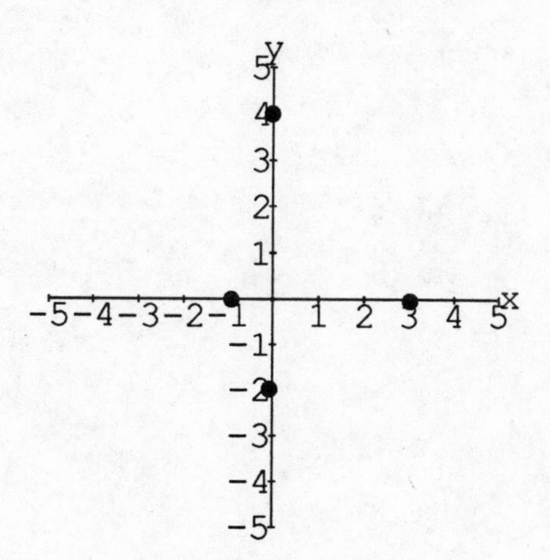

3.(a)

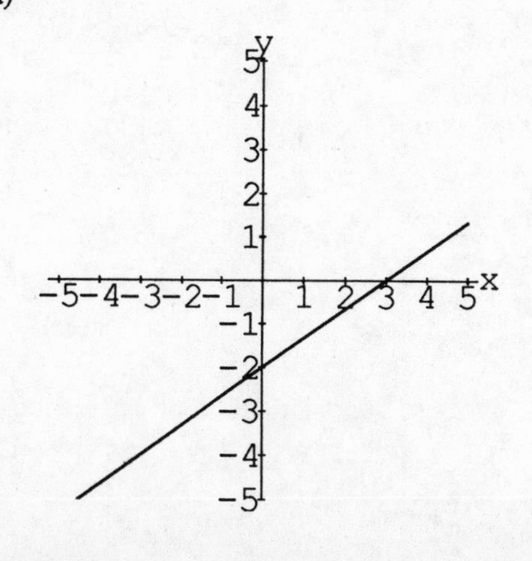

(b)

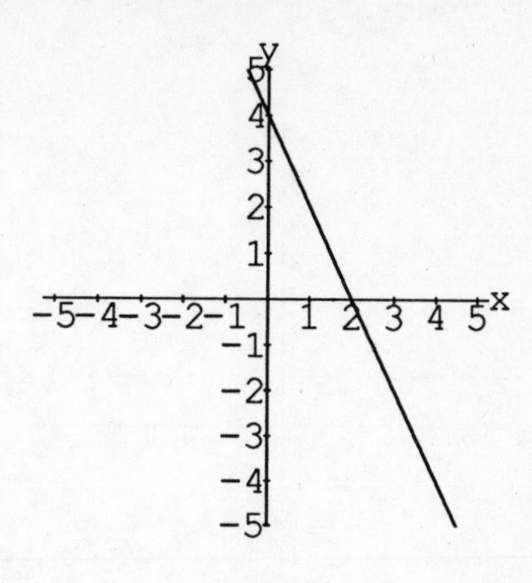

(c)

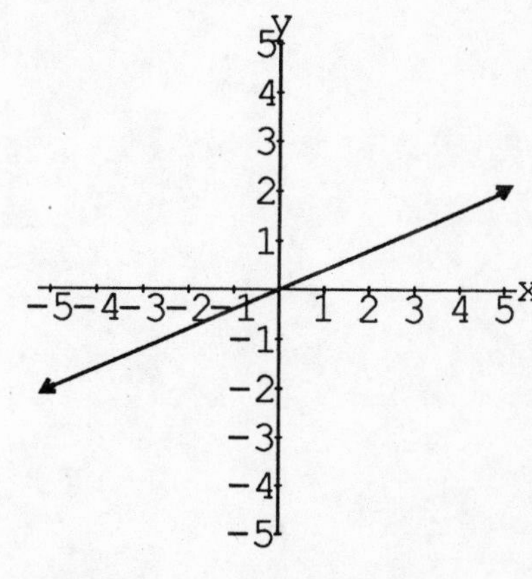

4.(a) x-intercept 4, y-intercept 2

(b) x-intercept 2, y-intercept –6

5.(a)

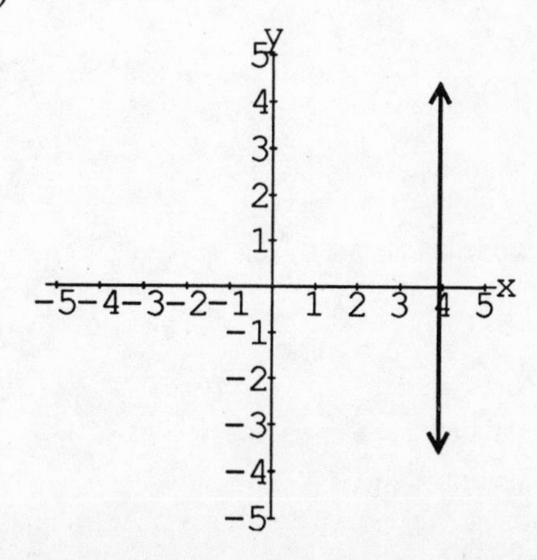

(b)

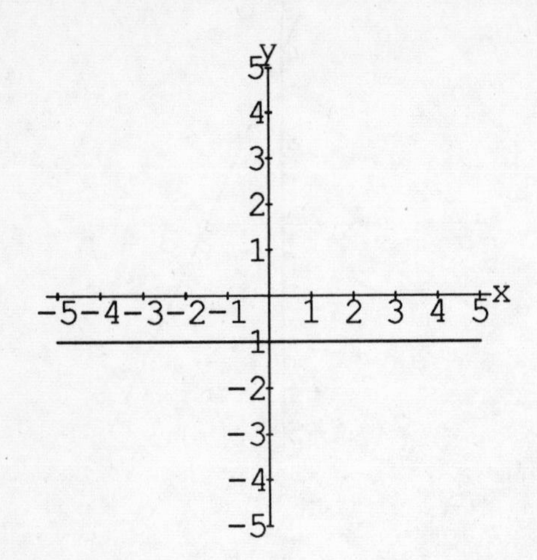

6.(a) $\sqrt{89}$

(b) $5\sqrt{5}$

7.(a) $(2, \frac{3}{2})$

(b) $((-\frac{5}{2}), 5)$

8.

(a) $\frac{3}{2}$

(b) $\frac{1}{6}$

(c) 0

(d) undefined

9.

(a) parallel

(b) perpendicular

(c) neither

10.

(a) $3x + y = 2$

(b) $x - 2y = -8$

(c) $4x + 7y = -26$

11.

(a) $y = -\frac{2}{3}x + 6$

(b) $y = 4x + 9$

12.

(a) $y = 2$

(b) $x = -4$

13.

(a) $y = -3x + 2$

(b) $y = \frac{1}{3}x - 3$

6

Rational Exponents and Roots

6.1 RATIONAL EXPONENTS

In this section we begin to study the relationship between rational exponents and radicals. The following list of roots will help you with this chapter.

Square Roots	Cube Roots	Fourth Roots	Fifth Roots
$\sqrt{1} = 1$	$\sqrt[3]{1} = 1$	$\sqrt[4]{1} = 1$	$\sqrt[5]{1} = 1$
$\sqrt{4} = 2$	$\sqrt[3]{8} = 2$	$\sqrt[4]{16} = 2$	$\sqrt[5]{32} = 2$
$\sqrt{9} = 3$	$\sqrt[3]{27} = 3$	$\sqrt[4]{81} = 3$	$\sqrt[5]{243} = 3$
$\sqrt{16} = 4$	$\sqrt[3]{64} = 4$	$\sqrt[4]{256} = 4$	$\sqrt[5]{1024} = 4$
$\sqrt{25} = 5$	$\sqrt[3]{125} = 5$	$\sqrt[4]{625} = 5$	$\sqrt[5]{3125} = 5$

n^{th} root

We know that the square root of 25 is 5 and the cube root of 8 is 2. We can write these statements using rational exponents as:

$$25^{1/2} = 5 \text{ and } 8^{1/3} = 2$$

This idea can be generalized:

$\sqrt[n]{b} = b^{1/n}$ for b a nonnegative number, and n a positive integer.

$b^{1/n}$ is read "the nth root of b."

EXAMPLE 1 Find each root.

a) $625^{1/4}$

b) $32^{1/5}$

c) $9^{1/2}$

SOLUTION 1

a) $625^{1/4} = 5$ since $\sqrt[4]{625} = 5$.
b) $32^{1/5} = 2$ since $\sqrt[5]{32} = 2$.
c) $9^{1/2} = 3$ since $\sqrt{9} = 3$.

Note that the even root of a negative number is not defined, so that $(-25)^{1/2}$ is *not* a real number. We *can* find the odd root of a negative number. Since $(-2)(-2)(-2) = -8$, we can say that $\sqrt[3]{-8} = -2$.

EXAMPLE 2 Find each root.

a) $(-32)^{1/5}$

b) $(-27)^{1/3}$

c) $(-125)^{1/3}$

d) $(-81)^{1/4}$

SOLUTION 2

a) $(-32)^{1/5} = -2$ since $(-2)^5 = -32$.
b) $(-27)^{1/3} = -3$ since $(-3)^3 = -27$.
c) $(-125)^{1/3} = -5$ since $(-5)^3 = -125$.

d) $(-81)^{1/4}$ is not a real number since the even root of a negative number is not a real number.

Definition of $a^{m/n}$

We can now expand our definition of $a^{1/n}$ to $a^{m/n}$:

If $a^{1/n}$ is a real number, then $a^{m/n} = (a^{1/n})^m$.

This definition tells us to first find the nth root of a, and then raise that answer to the m power.

EXAMPLE 3 Simplify.

a) $32^{3/5}$

b) $(-27)^{2/3}$

c) $9^{3/2}$

d) $(-16)^{3/4}$

SOLUTION 3

a) $32^{3/5} = (32^{1/5})^3 = 2^3 = 8$

b) $(-27)^{2/3} = [(-27)^{1/3}]^2 = (-3)^2 = 9$

c) $9^{3/2} = (9^{1/2})^3 = 3^3 = 27$

d) $(-16)^{3/4} = [(-16)^{1/4}]^3$ is not a real number since $(-16)^{1/4}$ is not a real number.

Exponent laws

We can now restate the exponents laws for rational number exponents.

Exponent Laws for a and b real numbers and m and n rational numbers:

$a^m \cdot a^n = a^{m+n}$ $\qquad$ $(a^m)^n = a^{mn}$

$\dfrac{a^m}{a^n} = a^{m-n}$ $\qquad$ $\left(\dfrac{a}{b}\right)^m = \dfrac{a^m}{b^m}$

$(ab)^m = a^m b^m$ $\qquad$ $a^{-m} = \dfrac{1}{a^m}$

EXAMPLE 4 Simplify. Write all answers with only positive exponents.

a) $2^{1/3} \cdot 2^{1/4}$

b) $(2^{1/3})^{1/4}$

c) $\dfrac{3^{3/4}}{3^{2/5}}$

d) $(8x^6y^3)^{2/3}$

e) $(\frac{27}{125})^{2/3}$

f) $64^{-5/6}$

SOLUTION 4

a) $2^{1/3} \cdot 2^{1/4} = 2^{1/3 + 1/4}$

To multiply, keep the bas the same and add the exponents.

$= 2^{4/12 + 3/12}$

$\frac{1}{3} \cdot \frac{1}{4} + \frac{1}{4} \cdot \frac{1}{3} = \frac{4}{12} + \frac{3}{12}.$

$= 2^{7/12}$

b) $(2^{1/3})^{1/4} = 2^{1/3 \cdot 1/4}$

To raise a power to a power, multiply the exponents.

$= 2^{1/12}$

$\frac{1}{3} \cdot \frac{1}{4} = \frac{1 \cdot 1}{3 \cdot 4} = \frac{1}{12}.$

c) $\dfrac{3^{3/4}}{3^{2/5}} = 3^{3/4 - 2/5}$

To divide, keep the base the same and subtract the exponents.

$= 3^{15/20 - 8/20}$

$(\frac{3}{4} \cdot \frac{1}{4}) - (\frac{2}{5} \cdot \frac{4}{4}) = \frac{15}{20} - \frac{8}{20} = \frac{7}{20}$

$= 3^{7/20}$

d) $(8x^6y^3)^{2/3} = 8^{2/3}\, x^{6 \cdot 2/3}\, y^{3 \cdot 2/3}$

To raise a product to a power, raise each factor to the power.

$= (8^{1/3})^2\, x^4\, y^2$

Use $a^{m/n} = (a^{1/n})^m$.

$= (2)^2 x^2 y^2$

$= 4x^2 y^2$

e) $\left(\frac{27}{125}\right)^{2/3} = \frac{27^{2/3}}{125^{2/3}}$ Use $\left(\frac{a}{b}\right)^m = \frac{a^m}{b^m}$.

$= \frac{(27^{1/3})^2}{(125^{1/3})^2}$ Use $a^{m/n} = (a^{1/n})^m$.

$= \frac{(3)^2}{(5)^2}$

$= \frac{9}{25}$

f) $64^{-5/6} = \frac{1}{64^{5/6}}$ Use $a^{-m} = \frac{1}{a^m}$.

$= \frac{1}{(64^{1/6})^5}$ Use $a^{m/n} = (a^{1/n})^m$.

$= \frac{1}{(2)^5}$

$= \frac{1}{32}$

EXAMPLE 5 Simplify. Write all answers with only positive exponents.

a) $p^{2/3} \cdot p^{-5/3}$

b) $\frac{(q^{3/4})^2}{(q^{1/4})^3}$

c) $(16r^8s^2)^{-3/4}$

d) $\left(\frac{x^{-3/5}}{y^{-2/5}}\right)^2 (x^{-1/5}y^{-3/5})^{-1}$

SOLUTION 5

a) $p^{2/3} \cdot p^{-5/3} = p^{2/3 + (-5/3)}$ — Use $a^m \cdot a^n = a^{m+n}$.

$= p^{-3/3}$ — $\frac{2}{3} + (-\frac{5}{3}) = \frac{2 + (-5)}{3} = \frac{-3}{3}$.

$= p^{-1}$

$= \frac{1}{p}$ — Use $a^{-m} = \frac{1}{a^m}$.

b) $\frac{(q^{3/4})^2}{(q^{1/4})^3} = \frac{q^{6/4}}{q^{3/4}}$ — Use $(a^m)^n = a^{m \cdot n}$.

$= q^{6/4 - 3/4}$ — Use $\frac{a^m}{a^n} = a^{m-n}$.

$= q^{3/4}$ — $\frac{6}{4} - \frac{3}{4} = \frac{6-3}{4} = \frac{3}{4}$.

c) $(16r^8s^2)^{-3/4} = \frac{1}{(16r^8s^2)^{3/4}}$ — Use $a^{-m} = \frac{1}{a^m}$.

$= \frac{1}{16^{3/4} r^{8 \cdot \frac{3}{4}} s^{2 \cdot \frac{3}{4}}}$ — Use $(a^m)^n = a^{m \cdot n}$.

$= \frac{1}{(16^{1/4})^3 r^6 s^{3/2}}$ — Use $a^{m/n} = (a^{1/n})^m$.

$= \frac{1}{(2)^3 r^6 s^{3/2}}$

$= \frac{1}{8r^6 s^{3/2}}$

d) $\left(\frac{x^{-3/5}}{y^{-2/5}}\right)^2 (x^{-1/5}y^{-3/5})^{-1}$

$= \frac{x^{-6/5}}{y^{-4/5}} (x^{1/5}y^{3/5})$ — Multiply each exponent in theparentheses by the exponent outside the parentheses.

$= x^{-6/5 + 1/5} y^{3/5 - (-4/5)}$

Add the exponents and keep the base for multiplication. Subtract the exponents and keep the base for division.

$= x^{-5/5} y^{7/5}$

Simplify the exponents.

$= x^{-1} y^{7/5}$

$= \dfrac{y^{7/5}}{x}$

Use $a^{-m} = \dfrac{1}{a^m}$.

6.2 RADICALS

Symbols and terms

The nth root of a in the previous section was written $a^{1/n}$. In this section, we will write

$$\sqrt[n]{a}$$

to represent the nth root of a, where n is called the **root** or **index**, a is called the **radicand** and the symbol $\sqrt{\ }$ is called a **radical sign.** An index of 2 (or square root) is not usually written, while all other indices (plural of index) must be written.

EXAMPLE 6 Find each root.

a) $\sqrt{49}$

b) $-\sqrt{49}$

c) $\sqrt[5]{32}$

d) $\sqrt[4]{625}$

e) $\sqrt[3]{-27}$

f) $\sqrt[4]{-81}$

SOLUTION 6

a) $\sqrt{49} = 7$ — Since no index is written, this is a square root.

b) $-\sqrt{49} = -1 \cdot \sqrt{49}$
$= -1 \cdot 7$
$= -7$

c) $\sqrt[5]{32} = 2$ — since $2^5 = 32$.

d) $\sqrt[4]{625} = 5$ — since $5^4 = 625$.

e) $\sqrt[3]{-27} = -3$ — since $(-3)^3 = -27$.

f) $\sqrt[4]{-81}$ is not a real number since the even root of a negative number is not a real number.

Rewriting $a^{m/n}$ with radicals

Since $a^{m/n} = (a^{1/n})^m$, and $a^{1/n} = \sqrt[n]{a}$, we can define $a^{m/n}$ using radicals.

If $a^{1/n}$ is a real number, then $a^{m/n} = (\sqrt[n]{a})^m = \sqrt[n]{a^m}$.

EXAMPLE 7 Rewrite each of the following with radicals.

a) $7^{1/2}$

b) $5^{2/3}$

c) $8x^{3/4}$

d) $p^{-1/3}$

SOLUTION 7

a) $7^{1/2} = \sqrt[2]{7^1} = \sqrt{7}$ — An index of 2 does not have to be written.

b) $5^{2/3} = \sqrt[3]{5^2} = \sqrt[3]{25}$

c) $8x^{3/4} = 8\,(\sqrt[4]{x^3})$

The exponent only applies to x here.

d) $p^{-1/3} = \dfrac{1}{p^{1/3}}$

Use $a^{-m} = \dfrac{1}{a^m}$.

$= \dfrac{1}{\sqrt[3]{p}}$

EXAMPLE 8 Rewrite each of the following with rational exponents.

a) $\sqrt{14}$

b) $\sqrt[5]{x^2}$

c) $(\sqrt[3]{p})^5$

d) $\dfrac{\sqrt{y^3}}{\sqrt[3]{y}}$

e) $\sqrt{\sqrt[3]{x}}$

SOLUTION 8

a) $\sqrt{14} = 14^{1/2}$

No written index means an index of 2.

b) $\sqrt[5]{x^2} = x^{2/5}$

The index becomes the denominator of the rational exponent.

c) $(\sqrt[3]{p})^5 = p^{5/3}$

The numerator of the rational exponent can be written under the radical, or outside the radical.

d) $\dfrac{\sqrt{y^3}}{\sqrt[3]{y}} = \dfrac{y^{3/2}}{y^{1/3}}$

Rewrite each radical using rational exponents.

$= y^{3/2 - 1/3}$

Use $\dfrac{a^m}{a^n} = a^{m-n}$.

$= y^{9/6 - 2/6}$ $\qquad (\frac{3}{2} \cdot \frac{3}{3}) - (\frac{1}{3} \cdot \frac{2}{2}) = \frac{9}{6} - \frac{2}{6}.$

$= y^{7/6}$

e) $\sqrt{\sqrt[3]{x}} = x^{1/3}$ $\qquad$ Rewrite $\sqrt[3]{x^1} = x^{1/3}$.

$= (x^{1/3})^{1/2}$ $\qquad$ No written index means an index of 2.

$= x^{1/3 \cdot 1/2}$ $\qquad$ Use $(a^m)^n = a^{m \cdot n}$.

$= x^{1/6}$

6.3 SIMPLIFYING RADICALS

The product rule

We will use the product rule to multiply radicals with the same index.

For a, b nonnegative real numbers, $\sqrt[n]{a} \cdot \sqrt[n]{b} = \sqrt[n]{ab}$.

EXAMPLE 9 Multiply.

a) $\sqrt{x} \cdot \sqrt{y}$

b) $\sqrt[3]{2} \cdot \sqrt[3]{x}$

c) $\sqrt{5} \cdot \sqrt{3xy}$

SOLUTION 9

a) $\sqrt{x} \cdot \sqrt{y} = \sqrt{xy}$ $\qquad$ Keep the index, 2, multiply the radicands.

b) $\sqrt[3]{2} \cdot \sqrt[3]{x} = \sqrt[3]{2x}$ $\qquad$ Keep the index, 3, multiply the radicands.

c) $\sqrt{5} \cdot \sqrt{3xy} = \sqrt{15xy}$ — Keep the index, 2, multiply the radicands.

The quotient rule

To begin our work with fractions involving radicals, we will use the quotient rule.

> For a, b, nonnegative real numbers, $\sqrt[n]{\dfrac{a}{b}} = \dfrac{\sqrt[n]{a}}{\sqrt[n]{b}}$, $b \neq 0$.

EXAMPLE 10 Simplify.

a) $\sqrt{\dfrac{16}{49}}$

b) $\sqrt{\dfrac{5}{81}}$

c) $\sqrt[3]{\dfrac{125}{8}}$

d) $\sqrt[3]{\dfrac{6}{27}}$

e) $\sqrt{\dfrac{x}{36}}$

SOLUTION 10

a) $\sqrt{\dfrac{16}{49}} = \dfrac{\sqrt{6}}{\sqrt{49}}$ — Use $\sqrt{\dfrac{a}{b}} = \dfrac{\sqrt{a}}{\sqrt{b}}$.

$= \dfrac{4}{7}$ — Simplify each radical.

b) $\sqrt{\dfrac{5}{81}} = \dfrac{\sqrt{5}}{\sqrt{81}}$ — Use $\sqrt{\dfrac{a}{b}} = \dfrac{\sqrt{a}}{\sqrt{b}}$.

$= \dfrac{\sqrt{5}}{9}$ — Simplify $\sqrt{81} = 9$.

c) $\sqrt[3]{\frac{125}{8}} = \frac{\sqrt[3]{125}}{\sqrt[3]{8}}$ Use $\sqrt[3]{\frac{a}{b}} = \frac{\sqrt[3]{a}}{\sqrt[3]{b}}$.

$= \frac{5}{2}$ Simplify each radical.

d) $\sqrt[3]{\frac{6}{27}} = \frac{\sqrt[3]{6}}{\sqrt[3]{27}}$ Use $\sqrt[3]{\frac{a}{b}} = \frac{\sqrt[3]{a}}{\sqrt[3]{b}}$.

$= \frac{\sqrt[3]{6}}{3}$ Simplify $\sqrt[3]{27} = 3$.

e) $\sqrt{\frac{x}{36}} = \frac{\sqrt{x}}{\sqrt{36}}$ Use $\sqrt{\frac{a}{b}} = \frac{\sqrt{a}}{\sqrt{b}}$.

$= \frac{\sqrt{x}}{6}$ Simplify $\sqrt{36} = 6$.

Simplifying $\sqrt[n]{x}$

We can use the product rule and the fact that $\sqrt[n]{x^n} = x$ to simplify radicals.

> **To simplify** $\sqrt[n]{x}$
> 1. Prime factor the radicand.
> 2. Write each factor to the nth power whenever possible.
> 3. Use $\sqrt[n]{x^n} = x$ to bring perfect factors outside the radical. Any factor raised to a power less than n stays in the radicand.

Note: We will assume all variables represent nonnegative real numbers.

EXAMPLE 11 Simplify.

a) $\sqrt{48x^4}$

b) $\sqrt[3]{48x^4}$

c) $\sqrt[4]{48x^4}$

d) $\sqrt[5]{48x^4}$

SOLUTION 11

a) $\sqrt{48x^4} = \sqrt{2^4 \cdot 3^1 \cdot x^4}$ — Factor the radicand.

$= \sqrt{2^2 \cdot 2^2 \cdot 3^1 \cdot x^2 \cdot x^2}$ — Write each base using the index, 2, as exponent whenever possible.

$= \sqrt{2^2}\sqrt{2^2}\sqrt{3^1}\sqrt{x^2}\sqrt{x^2}$ — $\sqrt{x \cdot y} = \sqrt{x}\sqrt{y}$

$= 2 \cdot 2 \cdot x \cdot x\sqrt{3}$ — Use $\sqrt[n]{x^n} = x$. Remember that no written index is an understood 2.

$= 4x^2\sqrt{3}$

b) $\sqrt[3]{48x^4} = \sqrt[3]{2^4 \cdot 3^1 \cdot x^4}$ — Factor the radicand.

$= \sqrt{2^3 \cdot 2^1 \cdot 3^1 \cdot x^3 \cdot x^1}$ — Write each base using the index, 3, as exponent.

$= \sqrt[3]{2^3}\sqrt[3]{2}\sqrt[3]{3}\sqrt[3]{x^3}\sqrt[3]{x}$ — $\sqrt[3]{xy} = \sqrt[3]{x}\sqrt[3]{y}$.

$= 2 \cdot x\sqrt[3]{2 \cdot 3 \cdot x}$ — Use $\sqrt[n]{x^n} = x$.

$\sqrt[3]{2^3} = 2$. $\sqrt[3]{x^3} = x$.

$= 2x\sqrt[3]{6x}$ — Multiply.

c) $\sqrt[4]{48x^4} = \sqrt[4]{2^4 \cdot 3^1 \cdot x^4}$ — Factor the radicand.

$= \sqrt[4]{2^4}\sqrt[4]{3^1}\sqrt[4]{x^4}$ — $\sqrt[4]{x \cdot y} = \sqrt[4]{x}\sqrt[4]{y}$.

$= 2 \cdot x\sqrt[4]{3}$ — Use $\sqrt[n]{x^n} = x$.

$= 2x\sqrt[4]{3}$ — Multiply.

d) $\sqrt[5]{48x^4} = \sqrt[5]{2^4 \cdot 3^1 \cdot x^4}$ — The index is greater than th exponents in the radicand.

$= \sqrt[5]{48x^4}$ This is already simplified.

A shortcut you may find helpful when simplifying radicals makes use of the definition $\sqrt[n]{a^m} = a^{m/n}$. Since $\frac{m}{n}$ is a division problem, we divide m by n, write the quotient as the exponent on a *outside* the radical, and the remainder as the exponent on a *inside* the radical. For example, to simplify $\sqrt[3]{y^{26}}$, divide 26 by 3:

$$\begin{array}{r} 8 \\ 3\overline{)26} \\ 24 \\ \hline 2 \end{array}$$

8 ← Exponent on y outside the radical

2 ← Exponent on y inside the radical

so $\sqrt[3]{y^{26}} = y^8\sqrt[3]{y^2}$.

EXAMPLE 12 Simplify.

a) $\sqrt[3]{a^5b^{19}c^{15}}$

b) $\sqrt[4]{32x^{15}y^{33}}$

SOLUTION 12

a) $\sqrt[3]{a^5b^{19}c^{15}}$ Divide the index into each exponent.

$$\begin{array}{r} 1 \\ 3\overline{)5} \\ 3 \\ \hline 2 \end{array} \qquad \begin{array}{r} 6 \\ 3\overline{)19} \\ 18 \\ \hline 1 \end{array} \qquad \begin{array}{r} 5 \\ 3\overline{)15} \\ 15 \\ \hline 0 \end{array}$$

$ab^6c^5\sqrt[3]{a^2b^1c^0}$ Remainders become exponents inside the radical.

$ab^6c^5\sqrt[3]{a^2b}$ Since $c^0 = 1$, $a^2bc^0 = a^2b(1) = a^2b$.

b) $\sqrt[4]{32x^{15}y^{33}} = \sqrt[4]{2^5x^{15}y^{33}}$ Write 32 as 2^5.

$$\begin{array}{r} 1 \\ 4\overline{)5} \\ 4 \\ 1 \end{array} \qquad \begin{array}{r} 3 \\ 4\overline{)15} \\ 12 \\ 3 \end{array} \qquad \begin{array}{r} 8 \\ 4\overline{)33} \\ 32 \\ 1 \end{array}$$

Divide the index into

each exponent.

$$= 2^1 x^3 y^8 \sqrt[4]{2^1 x^3 y^1}$$

Quotients are the exponents outside the radical.
Remainders are the exponents inside the radical.

$$= 2x^3 y^8 \sqrt[4]{2x^3 y}$$

Simplified form

The previous two examples simplified radicals so that no exponent in the radicand was greater than or equal to the index. There are other rules, summarized below, for determining when a radical is in simplified form.

A radical is in simplified form when:
1. The index is greater than any exponent in the radicand.
2. There are no fractions under the radical.
3. There are no radicals in the denominator of a fraction.

Although we have already worked with #1 and #2 from this list, we now turn our attention to #3. When the denominator of a fraction contains a radical that cannot be simplified to a rational number, we use the technique called "rationalizing the denominator." To rationalize a denominator that is a monomial, multiply the numerator and denominator by a radical that will make the denominator a rational number. (Binomial denominators will be treated in Section 6.5).

EXAMPLE 13 Simplify.

a) $\dfrac{3}{\sqrt{5}}$

b) $\dfrac{3\sqrt{5x}}{\sqrt{6y}}$

c) $\sqrt[3]{\dfrac{7}{2}}$

d) $\sqrt[3]{\dfrac{3x}{4y^2}}$

SOLUTION 13

a) $\frac{3}{\sqrt{5}} \cdot \frac{\sqrt{5}}{\sqrt{5}}$ — Multiply by $\frac{\sqrt{5}}{\sqrt{5}}$ to make the denominator a rational number.

$= \frac{3\sqrt{5}}{\sqrt{5^2}}$ — Use the product rule to multiply the radicals.

$= \frac{3\sqrt{5}}{5}$ — Simplify the denominator: $\sqrt{5^2} = 5$.

b) $\frac{3\sqrt{5x}}{\sqrt{6y}} \cdot \frac{\sqrt{6y}}{\sqrt{6y}}$ — Multiply by $\frac{\sqrt{6y}}{\sqrt{6y}}$ to make the denominator a rational number.

$= \frac{3\sqrt{5 \cdot 6 \cdot xy}}{\sqrt{6^2y^2}}$ — Multiply the radicals.

$= \frac{3\sqrt{30xy}}{6y}$ — Simplify the denominator.

$= \frac{\sqrt{30xy}}{2y}$ — Reduce $\frac{3}{6} = \frac{1}{2}$.

c) $\sqrt[3]{\frac{7}{2}} = \frac{\sqrt[3]{7}}{\sqrt[3]{2}}$ — Use the quotient rule to rewrite the radical.

$= \frac{\sqrt[3]{7}}{\sqrt[3]{2}} \cdot \frac{\sqrt[3]{2^2}}{\sqrt[3]{2^2}}$ — Multiplying by $\frac{\sqrt[3]{2^2}}{\sqrt[3]{2^2}}$ will not make the denominator a rational number since $\sqrt[3]{2} \cdot \sqrt[3]{2} = \sqrt[3]{4}$.

$= \frac{\sqrt[3]{7 \cdot 2^2}}{\sqrt[3]{2^3}}$ — Multiply the radicals.

$$= \frac{\sqrt[3]{28}}{2}$$ Simplify the denominator.

d) $$\sqrt[3]{\frac{3x}{4y^2}} = \frac{\sqrt{3x}}{\sqrt[3]{4y^2}}$$ Use the quotient rule to rewrite the radical.

$$= \frac{\sqrt[3]{3x}}{\sqrt[3]{2^2y^2}} \cdot \frac{\sqrt[3]{2y}}{\sqrt[3]{2y}}$$ Multiply by $\frac{\sqrt[3]{2y}}{\sqrt[3]{2y}}$ to make the denominator a rational number.

$$= \frac{\sqrt[3]{6xy}}{\sqrt[3]{2^3y^3}}$$ Multiply the radicals.

$$= \frac{\sqrt[3]{6xy}}{2y}$$ Simplify the denominator.

6.4 ADDING AND SUBTRACTING RADICAL EXPRESSIONS

Similar radicals

Only radicals with the same index and the same radicand (called **similar radicals**) can be added or subtracted. The distributive property allows us to add or subtract the numbers in front of similar radicals and keep the radicands the same:

$$5\sqrt{2} + 7\sqrt{2} = (5+7)\sqrt{2} = 12\sqrt{2}$$

If the radicands are not already simplified, you may need to simplify before (or after) combining similar radicals.

EXAMPLE 14 Simplify.

a) $3\sqrt{75}+\sqrt{27}$

b) $4\sqrt{125}-8\sqrt{20}+3\sqrt{45}$

c) $4y\sqrt{8x^2y^3}+7x\sqrt{72y^5}$

d) $\sqrt[3]{128a^3b^2}-6a\sqrt[3]{54b^2}$

SOLUTION 14

a) $3\sqrt{75}+\sqrt{27}$

$= 3\sqrt{5^2\cdot 3}+\sqrt{3^3}$ — Factor each radicand.

$= 3\cdot 5\sqrt{3}+3\sqrt{3}$ — Simplify each radical.

$= 15\sqrt{3}+3\sqrt{3}$

$= (15+3)\sqrt{3}$ — Use the distributive property.

$= 18\sqrt{3}$ — Simplify inside the parentheses.

b) $4\sqrt{125}-8\sqrt{20}+3\sqrt{45}$

$= 4\sqrt{5^3}-8\sqrt{2^2\cdot 5}+3\sqrt{3^2\cdot 5}$ — Factor each radicand.

$= 4\cdot 5\sqrt{5}-(8\cdot 2\sqrt{5})+3\cdot 3\sqrt{5}$ — Simplify each radical.

$= 20\sqrt{5}-16\sqrt{5}+9\sqrt{5}$

$= (20-6+9)\sqrt{5}$ — Use the distributive property.

$= 13\sqrt{5}$ — Simplify inside the parentheses.

c) $4y\sqrt{8x^2y^3}+7x\sqrt{72y^5}$

$= 4y\sqrt{2^3x^2y^3} + 7x\sqrt{2^3 \cdot 3^2y^5}$ Factor each radicand.

$= (4y) \cdot 2xy\sqrt{2y} + (7x) \cdot (2) \cdot 3y^2\sqrt{2y}$ Simplify each radical.

$= 8xy^2\sqrt{2y} + 42xy^2\sqrt{2y}$

$= (8xy^2 + 42xy^2)\sqrt{2y}$ Use the distributive property.

$= 50xy^2\sqrt{2y}$ Add similar terms in the parentheses.

d) $\sqrt[3]{128a^3b^2} - 6a\sqrt[3]{54b^2}$

$= \sqrt[3]{2^7a^3b^2} - 6a\sqrt[3]{2 \cdot 3^3b^2}$ Factor each radicand.

$= 2^2a\sqrt[3]{2b^2} - (6a) \cdot 3\sqrt[3]{2b^2}$ Simplify each radical.

$= 4a\sqrt[3]{2b^2} - 18a\sqrt[3]{2b^2}$

$= (4a - 18a)\sqrt[3]{2b^2}$ Use the distributive property.

$= -14a\sqrt[3]{2b^2}$ Add similar terms in the parentheses.

EXAMPLE 15 Simplify.

a) $\sqrt{72} - \dfrac{3}{\sqrt{2}}$

b) $3\sqrt{\dfrac{1}{2}} + \dfrac{\sqrt[3]{4}}{5}$

c) $\dfrac{2}{\sqrt{x}} - \sqrt{\dfrac{3}{x}}$

SOLUTION 15

a) $\sqrt{72} - \dfrac{3}{\sqrt{2}}$

$= \sqrt{2^3 \cdot 3^2} - \left(\dfrac{3}{\sqrt{2}} \cdot \dfrac{\sqrt{2}}{\sqrt{2}}\right)$ — Factor 72. Rationalize the denominator of $\dfrac{3}{\sqrt{2}}$.

$= 2 \cdot 3\sqrt{2} - \dfrac{3\sqrt{2}}{2}$ — Simplify each radical.

$= 6\sqrt{2} - \dfrac{3\sqrt{2}}{2}$ — Simplify the denominator.

$= (6 - \frac{3}{2})\sqrt{2}$ — Use the distributive property.

$= (\frac{12}{2} - \frac{3}{2})\sqrt{2}$ — Simplify inside the parentheses by subtracting fractions.

$= \frac{9}{2}\sqrt{2}$

b) $\sqrt[3]{\dfrac{1}{2}} + \dfrac{\sqrt[3]{4}}{5}$

$= \dfrac{\sqrt[3]{1}}{\sqrt[3]{2}} + \dfrac{\sqrt[3]{4}}{5}$ — Use the quotient rule: $\sqrt[3]{\dfrac{1}{2}} = \dfrac{\sqrt[3]{1}}{\sqrt[3]{2}}$

$= \dfrac{\sqrt[3]{1}}{\sqrt[3]{2}} \cdot \dfrac{\sqrt[3]{2^2}}{\sqrt[3]{2^2}} + \dfrac{\sqrt[3]{4}}{5}$ — Rationalize the denominator by multiplying by $\dfrac{\sqrt[3]{2^2}}{\sqrt[3]{2^2}}$.

$= \dfrac{\sqrt[3]{4}}{\sqrt[3]{2^3}} + \dfrac{\sqrt[3]{4}}{5}$ — Multiply the radicals.

$= \dfrac{\sqrt[3]{4}}{2} + \dfrac{\sqrt[3]{4}}{5}$ — Simplify the denominator.

$= (\frac{1}{2} + \frac{1}{5})\sqrt[3]{4}$ — Use the distributive property.

$= (\frac{5}{10} + \frac{2}{10}) \sqrt[3]{4}$ — Use a common denominator to add the fractions.

$= \frac{7}{10} \sqrt[3]{4}$

c) $\frac{2}{\sqrt{x}} - \sqrt{\frac{3}{x}}$

$= \frac{2}{\sqrt{3}} - \frac{\sqrt{3}}{\sqrt{x}}$ — Use the quotient rule: $\sqrt{\frac{3}{x}} = \frac{\sqrt{3}}{\sqrt{x}}$.

$= \left(\frac{2}{\sqrt{x}} \cdot \frac{\sqrt{x}}{\sqrt{x}}\right) - \left(\frac{\sqrt{3}}{\sqrt{x}} \cdot \frac{\sqrt{x}}{\sqrt{x}}\right)$ — Rationalize each denominator.

$= \frac{2\sqrt{x}}{\sqrt{x^2}} - \frac{\sqrt{3x}}{\sqrt{x^2}}$ — Multiply the radicals.]

$= \frac{2\sqrt{x}}{x} - \frac{\sqrt{3x}}{x}$ — Simplify each denominator.

$= (2\sqrt{x} - \sqrt{3x})\,\frac{1}{x}$ — Use the distributive property.

The terms inside the parentheses are not similar radicals and cannot be combined.

6.5MULTIPLYING AND DIVIDING RADICAL EXPRESSIONS

Multiplying radicals

We use the commutative and associative properties to multiply radicals with equal indices:

$$(a\sqrt[n]{x})\,(b\sqrt[n]{y}) =$$

$$(a \cdot b)\sqrt[n]{x}\sqrt[n]{y} =$$

$$ab\sqrt[n]{xy}$$

In other words, to multiply radicals with the same index, multiply the numbers outside the radical, multiply the radicands, and keep the index the same. We use the distributive property to multiply radical expressions involving more than monomials.

EXAMPLE 16 Multiply.

a) $(4\sqrt{2})\,(6\sqrt{5})$

b) $\sqrt{5}\,(2\sqrt{3} - 4\sqrt{15})$

c) $(3\sqrt{7} - \sqrt{5})\,(2\sqrt{7} + 3\sqrt{5})$

d) $(\sqrt{6} - 4)^2$

e) $(\sqrt{x} + \sqrt{y})\,(\sqrt{x} - \sqrt{y})$

SOLUTION 16

a) $(4\sqrt{2})\,(6\sqrt{5}) = (4 \cdot 6)\,\sqrt{2}\sqrt{5}$ — Multiply the numbers outside the radicals.

$= 24\sqrt{10}$ — Multiply the radicands.

b) $\sqrt{5}\,(2\sqrt{3} - 4\sqrt{15})$

$= \sqrt{5} \cdot 2\sqrt{3} - (\sqrt{5} \cdot 4\sqrt{15})$ — Use the distributive property.

$= 2\sqrt{15} - 4\sqrt{75}$ — Multiply each term.

$= 2\sqrt{15} - 4\sqrt{5^2 \cdot 3}$ — Factor the radicand.

$= 2\sqrt{15} - (4 \cdot 5\sqrt{3})$ — Simplify the radical.

$= 2\sqrt{15} - 20\sqrt{3}$ — Stop here. Only similar radicals can be subtracted.

c) $(3\sqrt{7} - \sqrt{5})\,(2\sqrt{7} + 3\sqrt{5})$ — Use FOIL.

First Outer Inner Last

$$= ((3\sqrt{7}) \cdot (2\sqrt{7})) + ((3\sqrt{7}) \cdot (3\sqrt{5})) - (\sqrt{5} \cdot 2\sqrt{7}) - (\sqrt{5} \cdot 3\sqrt{5})$$

$= 6\sqrt{49} + 9\sqrt{35} - 2\sqrt{35} - 3\sqrt{25}$	Multiply each term.
$= 6\sqrt{7^2} + 9\sqrt{35} - 2\sqrt{35} - 3\sqrt{5^2}$	Simplify radicals.
$= 4 + 9\sqrt{35} - 2\sqrt{35} - 15$	
$= (42 - 15) + (9 - 2)\sqrt{35}$	Combine similar terms.
$= 27 + 7\sqrt{35}$	Stop here. Only similar radicals can be added.
d) $(\sqrt{6} - 4)^2$	We can use the rules for squaring a binomial.
$(\sqrt{6})^2 = 6$	Square the first term.
$2(\sqrt{6})(-4) = -8\sqrt{6}$	Multiply 2 times the first term $(\sqrt{6})$ times the last term (–4).
$(-4)^2 = 16$	Square the last term.
$= 6 - 8\sqrt{6} + 16$	Add the terms together.
$= 22 - 8\sqrt{6}$	Add similar terms: $6 + 16 = 22$.
e) $(\sqrt{x} + \sqrt{y})(\sqrt{x} - \sqrt{y})$	We can use the rules for multiplying a sum and difference of two terms.
$(\sqrt{x})^2 = x$	Square the first term.
$(\sqrt{y})^2 = y$	Square the last term.
$= x - y$	Subtract.

Dividing radicals

We have already used the quotient rule and rationalizing the denominator to divide radicals with monomial denominators. When the denominator contains a binomial with one or two radicals, we multiply by the **conjugate** of the denominator. The conjugate of $a + b$ is $a - b$, and the conjugate of $a - b$ is $a + b$.

EXAMPLE 17 Rationalize the denominator in the following expressions.

a) $\dfrac{6}{5-\sqrt{3}}$

b) $\dfrac{\sqrt{3}-2}{\sqrt{3}+2}$

c) $\dfrac{a}{\sqrt{a}+\sqrt{b}}$

SOLUTION 17

a) $\dfrac{6}{5-\sqrt{3}}$

$= \dfrac{6}{5-\sqrt{3}} \cdot \dfrac{5+\sqrt{3}}{5+\sqrt{3}}$ — The conjugate of $5-\sqrt{3}$ is $5+\sqrt{3}$.

$= \dfrac{6(5+\sqrt{3})}{(5-\sqrt{3})(5+\sqrt{3})}$ — Multiply the numerators and the denominators.

$= \dfrac{6(5+\sqrt{3})}{(5)^2-(\sqrt{3})^2}$ — Use the rules for multiplying a sum and difference of two terms.

$= \dfrac{6(5+\sqrt{3})}{25-3}$ — Square both terms in the denominator.

$= \dfrac{\overset{3}{\cancel{6}}(5+\sqrt{3})}{\underset{11}{\cancel{22}}}$ — Simplify the denominator.

$= \dfrac{3(5+\sqrt{3})}{11}$ — Reduce: $\dfrac{6}{22} = \dfrac{3}{11}$.

b) $\dfrac{\sqrt{3}-2}{\sqrt{3}+2}$

$= \dfrac{\sqrt{3}-2}{\sqrt{3}+2} \cdot \dfrac{\sqrt{3}-2}{\sqrt{3}-2}$ — The conjugate of $\sqrt{3}+2$ is $\sqrt{3}-2$.

$= \dfrac{(\sqrt{3})^2 + 2(\sqrt{3})(-2) + (-2)^2}{((\sqrt{3})^2 - (2)^2)}$ — Use the rules for squaring a binomial in the numerator.

$= \dfrac{3-4\sqrt{3}+4}{3-4}$ — Simplify the numerator and denominator.

$= \dfrac{7-4\sqrt{3}}{-1}$ — Combine similar terms.

$= -1(7-4\sqrt{3})$ — Simplify the fraction.

$= -7+4\sqrt{3}$ — Substitute the −1.

c) $\dfrac{a}{\sqrt{a}+\sqrt{b}}$

$= \dfrac{a}{\sqrt{a}+\sqrt{b}} \cdot \dfrac{\sqrt{a}-\sqrt{b}}{\sqrt{a}-\sqrt{b}}$ — The conjugate of $\sqrt{a}+\sqrt{b}$ is $\sqrt{a}-\sqrt{b}$.

$= \dfrac{a(\sqrt{a}-\sqrt{b})}{(\sqrt{a})^2 - (\sqrt{b})^2}$ — Use the rules for multiplying a sum and difference of two terms.

$= \dfrac{a(\sqrt{a}-\sqrt{b})}{a-b}$ — Square each term in the denominator.

$= \dfrac{a\sqrt{a}-a\sqrt{b}}{a-b}$ — Distribute a.

6.6 SOLVING EQUATIONS CONTAINING RADICALS

Solving equations containing radicals makes use of the technique of raising both sides of an equation to a power that will eliminate one radical at a time. This technique can produce extraneous solutions -- that is, solutions that will not check in the original equation. You *must* check every solution.

To Solve an Equation Containing Radicals:

1. Isolate one radical on one side of the equation.
2. Raise both sides of the equation to a power equal to the index (for square roots, square both sides; for cube roots, cube each side, etc.).
3. Combine similar terms.
4. If there is still a radical in the equation, go back to Step 1.
5. Solve the resulting equation. If the resulting equation is linear, isolate the variable. If the resulting equation is a quadratic equation, get 0 on one side and factor.
6. Check all proposed solutions in the original equation.

Equations containing one radical

When the equation contains one radical, you will only have to raise each side to a power one time.

EXAMPLE 18 Solve.

a) $\sqrt{2x-3} = 3$

b) $\sqrt{4s+9}+5 = 0$

c) $\sqrt[3]{5p-2} = 2$

d) $\sqrt{3q+1} = q-3$

SOLUTION 18

a) $\sqrt{2x-3} = 3$

$(\sqrt{2x-3})^2 = (3)^2$ The index is 2, so square both sides.

$2x-3=9$ This is a linear equation, so isolate x.

$2x-3+3=9+3$	Add 3 to both sides.
$2x=12$	Divide both sides by 2.
$x=6$	Proposed solution.

Check:

$\sqrt{2(6)-3} \stackrel{?}{=} 3$	Substitute $x=6$ into the original equation.
$\sqrt{9} \stackrel{?}{=} 3$	
$3=3$	True, so $x=6$ is the solution.

b) $\sqrt{4s+9}+5=0$

$\sqrt{4s+9}+5-5=0-5$	Isolate the radical by subtracting 5 from both sides.
$\sqrt{4s+9}=-5$	
$(\sqrt{4s+9})^2=(-5)^2$	The index is 2, so square both sides.
$4s+9=25$	This is a linear equation, so isolate s.
$4s+9-9=25-9$	Subtract 9 from both sides.
$4s=16$	Divide both sides by 4.
$s=4$	Proposed solution.

Check:

$\sqrt{4(4)+9}+5 \stackrel{?}{=} 0$	Substitute $s=4$ into the original equation.
$\sqrt{25}+5 \stackrel{?}{=} 0$	
$5+5 \stackrel{?}{=} 0$	
$10=0$	False.

Since the check is false, 4 is *not* a solution. Therefore the solution set is empty, $\varnothing$.

c) $\sqrt[3]{5p-2} = 2$

$(\sqrt[3]{5p-2})^3 = (2)^3$ — The index is 3, so cube both sides.

$5p - 2 = 8$ — This is a linear equation, so isolate *p*.

$5p - 2 + 2 = 8 + 2$ — Add 2 to both sides.

$5p = 10$ — Divide both sides by 5.

$p = 2$ — Proposed solution.

Check:

$\sqrt[3]{5(2) - 2} \stackrel{?}{=} 2$ — Substitute $p = 2$ into the original equation.

$\sqrt[3]{8} \stackrel{?}{=} 2$

$2 = 2$ — True, so $p = 2$ is the solution.

d) $\sqrt{3q+1} = q - 3$

$(\sqrt{3q+1})^2 = (q-3)^2$ — The index is 2, so square both sides.

$3q + 1 = q^2 - 6q + 9$ — $(q-3)^2 = (q)^2 + 2(q)(-3) + (-3)^2$.

$3q + 1 - 1 - 3q = q^2 - 6q + 9 - 3q - 1$ — This is a quadratic equation, so get 0 on one side.

$0 = q^2 - 9q + 8$

$(q-8)(q-1) = 0$ — Factor.

$q = 8$ or $q = 1$ — Set each factor equal to 0 and solve.

Check:

If $q = 8$

$\sqrt{3(8)+1} \stackrel{?}{=} (8) - 3$

$\sqrt{25} \stackrel{?}{=} 5$

If $q = 1$

$\sqrt{3(1)+1} \stackrel{?}{=} (1) - 3$

$\sqrt{4} \stackrel{?}{=} -2$

$5 = 5$ $\qquad$ $2 = -2$

Only $q = 8$ is a solution.

Equations containing two radicals

When the equation contains more than one radical, you may have to repeat the process of isolating a radical and raising both sides to a power more than one time.

EXAMPLE 19 Solve.

a) $\sqrt{x+16} = \sqrt{x}+2$

b) $\sqrt{2x+3} + \sqrt{4x+5} = 2$

c) $\sqrt[4]{2x-6} = \sqrt[4]{x-1}$

SOLUTION 19

a) $\sqrt{x+16} = \sqrt{x}+2$ — One radical is isolated.

$(\sqrt{x+16})^2 = (\sqrt{x}+2)^2$ — The index is 2, so square both sides.

$x+16 = x+4\sqrt{x}+4$ — $(\sqrt{x}+2)^2 =$

$(\sqrt{x})^2 + 2(\sqrt{x})(2) + (2)^2$

$x+6-x-4 = x+4\sqrt{x}+4-x-4$ — There is still a radical in the equation. Isolate the radical.

$12 = 4\sqrt{x}$ — Divide both sides by 4.

$3 = \sqrt{x}$

$(3)^2 = (\sqrt{x})^2$ — The index is 2, so square both sides.

$9 = x$ — Proposed solution.

Check:

$\sqrt{9+16} \stackrel{?}{=} \sqrt{9}+2$ — Substitute $x = 9$ into the original equation.

$\sqrt{25} \stackrel{?}{=} 3+2$

$5 = 5$ — True, so $x = 9$ is the solution.

b) $\sqrt{2x+3}+\sqrt{4x+5} = 2$

$\sqrt{2x+3}+\sqrt{4x+5}-\sqrt{4x+5} = 2-\sqrt{4x+5}$ Isolate $\sqrt{2x+3}$.

$\sqrt{2x+3} = 2-\sqrt{4x+5}$

$(\sqrt{2x+3})^2 = (2-\sqrt{4x+5})^2$ — The index is 2, so square both sides.

$2x+3 = 4-4\sqrt{4x+5}+4x+5$

$2x+3 = -4\sqrt{4x+5}+4x+9$ — Combine similar terms.

$2x+3-4x-9 = -4\sqrt{4x+5}+4x+9-4x-9$ Isolate the radical.

$-2x-6 = -4\sqrt{4x+5}$ — You could divide both sides by –4, but it is easier to square both sides.

$(-2x-6)^2 = (-4\sqrt{4x+5})^2$

$4x^2 + 24x + 36 = 16(4x+5)$ — Be careful to square –4.

$4x^2 + 24x + 36 = 64x + 80$ — Simplify the right sides.

$4x^2 - 40x - 44 = 0$ — Get 0 alone on one side.

$4(x^2 - 10x - 11) = 0$ — Factor.

$4(x-11)(x+1) = 0$ — Factor.

$x - 11 = 0$ or $x + 1 = 0$ — Set each factor equal to 0 and solve.

$x = 11$ or $x = -1$ — Proposed solutions.

Check:

If $x = 11$,

$$\sqrt{2(11)+3} + \sqrt{4(11)+5} \stackrel{?}{=} 2$$ Substitute $x = 11$ into the original equation.

$$\sqrt{25} + \sqrt{49} \stackrel{?}{=} 2$$

$$5 + 7 \stackrel{?}{=} 2$$

$$12 = 2$$ False.

If $x = -1$,

$$\sqrt{2(-1)+3} + \sqrt{4(-1)+5} \stackrel{?}{=} 2$$ Substitute $x = -1$ into the original equation.

$$\sqrt{1} + \sqrt{1} \stackrel{?}{=} 2$$

$$2 = 2$$ True.

There is only one solution to the original equation, $x = -1$.

c) $\sqrt[4]{2x-6} = \sqrt[4]{x-1}$ A radical is already isolated.

$$\left(\sqrt[4]{2x-6}\right)^4 = \left(\sqrt[4]{x-1}\right)^4$$ The index is 4, so raise each side to the fourth power.

$$2x - 6 = x - 1$$ This is a linear equation, so isolate x.

$$x = 5$$ Proposed solution.

Check:

$$\sqrt[4]{2(5)-6} \stackrel{?}{=} \sqrt[4]{5-1}$$ Substitute $x = 5$ into the original equation.

$$\sqrt[4]{4} = \sqrt[4]{4}$$ True, so $x = 5$ is the solution.

6.7 COMPLEX NUMBERS

There is no real number whose square is a negative number. We will now introduce a number system called the complex number system that defines a number i whose square *is* negative.

Imaginary numbers

We begin our work with a definition of i.

Definition. $i = \sqrt{-1}$ and $i^2 = -1$.

Numbers that are multiples of i are called imaginary numbers. Square roots with negative radicands can be simplified using –1 as a factor.

EXAMPLE 20 Simplify.

a) $\sqrt{-25}$

b) $\sqrt{-7}$

SOLUTION 20

a) $\sqrt{-25} = \sqrt{(-1) \cdot 25}$ Factor.

$= i\sqrt{25}$ $\sqrt{-1} = i$.

$= 5i$

b) $\sqrt{-7} = \sqrt{(-1) \cdot 7}$ Factor.

$= i\sqrt{7}$ $\sqrt{-1} = i$.

Complex numbers

By combining i with the real numbers, we form a new set of numbers called the complex numbers.

Definition. A complex number is any number that can be written in the form $a + bi$ where a and b are real numbers and $i = \sqrt{-1}$.

$a + bi$ is called the **standard form** of a complex number, where a is called the **real part** and b is called the **imaginary part**. When working with complex numbers, you can treat i like a variable except that powers of i must be simplified.

EXAMPLE 21 Simplify.

a) i^3

b) i^4

c) i^{24}

d) i^{18}

SOLUTION 21

a) $i^3 = i^2 \cdot i$	Factor i^3.
$= (-1)i$	Use $i^2 = -1$.
$= -i$	
b) $i^4 = i^2 \cdot i^2$	Factor i^4.
$= (-1)(-1)$	Use $i^2 = -1$.
$= 1$	
c) $i^{24} = (i^4)^6$	Use exponent laws.
$= (1)^6$	Use $i^4 = 1$.
$= 1$	
d) $i^{18} = (i^4)^4 \cdot i^2$	Use exponent laws.
$= (1)^4 \cdot (-1)$	Use $i^4 = 1$ and $i^2 = -1$.
$= -1$	

There are many ways to break down powers of i. It is often convenient to work with $i^4 = 1$ as shown in *c* and *d*. Notice that all powers of i can be simplified to $1, -1, i$ or $-i$.

Adding and subtracting complex numbers

Because our commutative and associative properties are valid for complex numbers, we can add and subtract complex numbers by treating i like a variable.

EXAMPLE 22 Add or subtract, as indicated.

a) $(5 + 3i) + (8 - 2i)$

b) $(3 - 7i) - (2 - 6i)$

c) $8 + (4 - 3i)$

d) $(2 + 8i) - (-6 - 3i)$

SOLUTION 22

a) $(5 + 3i) + (8 - 2i)$

$= (5 + 8) + (3 - 2)i$ — Combine the real parts and combine the imaginary parts.

$= 13 + i$

b) $(3 - 7i) - (2 - 6i)$

$= (3 - 2) + (-7 - (-6))i$ — Combine the real parts and combine the imaginary parts.

$= 1 + -1i$

$= 1 - i$

c) $8 + (4 - 3i)$

$= (8 + 4) - 3i$ — Combine the real parts.

$= 12 - 3i$

d) $(2 + 8i) - (-6 - 3i)$

$= (2 - (-6)) + (8 - (-3))i$ — Combine the real parts and combine the imaginary parts.

$= 8 + 11i$

Multiplying complex numbers

Because the distributive property is valid for complex numbers, we can multiply complex numbers in the usual manner with the additional step of simplifying any powers of *i*.

EXAMPLE 23 Multiply.

a) $(2 - 5i)(3 + 4i)$

b) $(1 + 4i)(1 - 4i)$

SOLUTION 23

a) $(2 - 5i)(3 + 4i)$

$= 2(3) + 2(4i) - 5i(3) - 5i(4i)$	Use FOIL.
$= 6 + 8i - 15i - 20i^2$	
$= 6 - 7i - 20i^2$	Combine similar terms.
$= 6 - 7i - 20(-1)$	Use $i^2 = -1$.
$= 6 - 7i + 20$	
$= 26 - 7i$	Combine similar terms.

b) $(1 + 4i)(1 - 4i)$

$= 1(1) + 1(-4i) + 4i(1) + 4i(-4i)$	Use FOIL.
$= 1 - 4i + 4i - 16i^2$	
$= 1 - 16i^2$	Combine similar terms.
$= 1 - 16(-1)$	Use $i^2 = -1$.
$= 1 + 16$	Add.
$= 17$	

Dividing complex numbers

If a complex number is written as a fraction that contains i in the denominator, we use a process similar to rationalizing the denominator for radicals to write that complex number in standard form.

To Write Complex Number Quotients in Standard Form

1. If the denominator is a monomial involving i, multiply by $\frac{i}{i}$ and simplify. Write the answer in the form $a + bi$.
2. If the denominator is a binomial involving i, multiply the numerator and denominator by the conjugate of the denominator.
3. Write the answer as $a + bi$.

EXAMPLE 24 Find the quotients.

a) $\frac{5 - 2i}{i}$

b) $\dfrac{2-3i}{1+4i}$

SOLUTION 24

a) $\dfrac{5-2i}{i} = \dfrac{5-2i}{i} \cdot \dfrac{i}{i}$ — Monomial denominator so multiply by $\dfrac{i}{i}$.

$= \dfrac{5i-2i^2}{i^2}$ — Multiply.

$= \dfrac{5i-2(-1)}{(-1)}$ — Use $i^2 = -1$.

$= \dfrac{5i+2}{-1}$

$= -5i-2$

$= -2-5i$ — Write the answer in standard form.

b) $\dfrac{2-3i}{1+4i} = \dfrac{2-3i}{1+4i} \cdot \dfrac{1-4i}{1-4i}$ — Binomial denominator so multiply by the conjugate.

$= \dfrac{2(1)+2(-4i)-3i(1)-3i(-4i)}{1(1)+1(-4i)+4i(1)+4i(-4i)}$ — Use FOIL.

$= \dfrac{2-8i-3i+12i^2}{1-4i+4i-16i^2}$ — Multiply.

$= \dfrac{2-11i+12i^2}{1-16i^2}$ — Combine similar terms.

$= \dfrac{2-11i+12(-1)}{1-16(-1)}$ — Use $i^2 = -1$.

$= \dfrac{2-11i-12}{1+16}$ — Multiply.

$= \dfrac{-10-11i}{17}$ — Combine similar terms.

$= -\dfrac{10}{17} - \dfrac{11}{17}i$ — Write the answer in standard form.

Practice Exercises

1. Find each root.

(a) $25^{1/2}$

(b) $81^{1/4}$

(c) $27^{1/3}$

(d) $(-64)^{1/3}$

(e) $(-243)^{1/5}$

2. Simplify.

(a) $16^{3/2}$

(b) $(-8)^{2/3}$

(c) $256^{3/4}$

(d) $(-81)^{3/2}$

3. Simplify. Write all answers with only positive exponents.

(a) $5^{1/2} \cdot 5^{1/3}$

(b) $(3^{1/2})^{1/4}$

(c) $\dfrac{7^{2/3}}{7^{1/4}}$

(d) $(25x^4y^8)^{3/2}$

(e) $\left(\dfrac{8}{125}\right)^{4/3}$

(f) $1024^{-3/5}$

4. Simplify. Write all answers with only positive exponents.

(a) $m^{2/7} \cdot m^{-4/7}$

(b) $\dfrac{(r^{5/6})^2}{(r^{1/6})^3}$

(c) $(64a^6b^9)^{-2/3}$

(d) $\left(\dfrac{x^{-2/7}}{y^{-3/7}}\right)^2 (x^{-3/7}y^{-4/7})^{-1}$

5. Find each root.

(a) $\sqrt{81}$

(b) $-\sqrt{81}$

(c) $\sqrt[5]{1}$

(d) $\sqrt[3]{-64}$

(e) $\sqrt{-9}$

6. Rewrite each of the following with radicals.

(a) $11^{1/2}$

(b) $6^{3/4}$

(c) $5x^{1/3}$

(d) $s^{-1/2}$

7. Rewrite each of the following with rational exponents.

(a) $\sqrt{19}$

(b) $\sqrt[3]{x^2}$

(c) $(\sqrt[4]{x})^2$

(d) $\dfrac{\sqrt{p^5}}{\sqrt[3]{p}}$

(e) $\sqrt[3]{\sqrt{x}}$

8. Multiply.

(a) $\sqrt{pq} \cdot \sqrt{r}$

(b) $\sqrt[3]{6} \cdot \sqrt[3]{y^2}$

(c) $\sqrt{2} \cdot \sqrt{7xy}$

9. Simplify.

(a) $\sqrt{\dfrac{25}{36}}$

(b) $\sqrt{\dfrac{7}{100}}$

(c) $\sqrt[3]{\dfrac{64}{125}}$

(d) $\sqrt[3]{\dfrac{9}{8}}$

(e) $\sqrt{\dfrac{x}{49}}$

10. Simplify.

(a) $\sqrt[3]{a^8 b^{12} c^{16}}$

(b) $\sqrt[4]{243x^{16} y^{19}}$

11. Simplify.

(a) $\dfrac{2}{\sqrt{7}}$

(b) $\dfrac{4\sqrt{11x}}{\sqrt{2y}}$

(c) $\sqrt[3]{\dfrac{5}{3}}$

(d) $\sqrt[3]{\dfrac{4x}{5y^2}}$

12. Simplify.

(a) $-4\sqrt{90} + \sqrt{40}$

(b) $4\sqrt{32} - \sqrt{18} + 5\sqrt{128}$

(c) $3a\sqrt{20a^3 b^2} + 2b\sqrt{45a^5}$

(d) $\sqrt[3]{54a^3 b^4} - 4a\sqrt[3]{16b^4}$

13. Simplify.

(a) $\sqrt{50} + \dfrac{5}{\sqrt{2}}$

(b) $\sqrt[3]{\dfrac{1}{3}} + \dfrac{\sqrt[3]{9}}{4}$

c) $\dfrac{4}{\sqrt{x}} - \sqrt{\dfrac{2}{x}}$

14. Multiply.

(a) $(3\sqrt{5})\,(4\sqrt{7})$

(b) $\sqrt{2}\,(5\sqrt{3} + 4\sqrt{2})$

(c) $(2\sqrt{3} - \sqrt{5})\,(6\sqrt{3} + 2\sqrt{5})$

(d) $(\sqrt{5} + 3)^2$

(e) $(\sqrt{a} + \sqrt{b})\,(\sqrt{a} - \sqrt{b})$

15. Rationalize the denominator in the following expressions.

(a) $\dfrac{4}{6 - \sqrt{2}}$

(b) $\dfrac{2 + \sqrt{5}}{2 - \sqrt{5}}$

(c) $\dfrac{x}{\sqrt{x} - \sqrt{y}}$

16. Solve.

(a) $\sqrt{2x - 1} = 3$

(b) $\sqrt{3p+1}+4 = 0$

(c) $\sqrt[3]{7p+6} = 3$

(d) $\sqrt{5y+29} = y+3$

e) $\sqrt{3a} = \sqrt{a+6}$

(f) $\sqrt{3x-2}-\sqrt{2x-8} = 2$

(g) $\sqrt[3]{y-7} = \sqrt[3]{2y+4}$

17. Simplify.

(a) $\sqrt{-49}$

(b) $\sqrt{-6}$

18. Simplify.

(a) i^5

(b) i^8

(c) i^{14}

(d) i^{27}

19. Add or subtract as indicated.

(a) $(4 + 2i) + (7 - 8i)$

(b) $(2 - 5i) - (6 + 10i)$

(c) $5 + (3 - 4i)$

(d) $2i - (-4 - 5i)$

20. Multiply.

(a) $(3 - 4i)(2 + 6i)$

(b) $(2 + 3i)(2 - 3i)$

21. Divide.

(a) $\dfrac{6+3i}{i}$

(b) $\dfrac{3+4i}{2-i}$

Solutions

1.(a) 5

(b) 3

(c) 3

(d) –4

(e) –3

(f) not a real number

2.(a) 64

(b) 4

(c) 64

(d) not a real number

3.(a) $5^{5/6}$

(b) $3^{1/8}$

(c) $7^{5/12}$

(d) $125x^6y^{12}$

(e) $\frac{16}{625}$

(f) $\frac{1}{64}$

4.(a) $\frac{1}{m^{2/7}}$

(b) $r^{7/6}$

(c) $\frac{1}{16a^4b^6}$

(d) $\frac{y^{10/7}}{x^{1/7}}$

5.(a) 9

(b) –9

(c) 1

(d) –4

(e) not a real solution

6.(a) $\sqrt{11}$

(b) $\sqrt[4]{6^3} = \sqrt[4]{216}$

(c) $5\sqrt[3]{x}$

(d) $\frac{1}{\sqrt{5}}$

7.(a) $19^{1/2}$

(b) $x^{2/3}$

(c) $x^{1/2}$

(d) $p^{13/6}$

(e) $x^{1/6}$

8.(a) $\sqrt{pqr}$

(b) $\sqrt[3]{6y^2}$

(c) $\sqrt{14xy}$

9.(a) $\frac{5}{6}$

(b) $\frac{\sqrt{7}}{10}$

(c) $\frac{4}{5}$

(d) $\frac{\sqrt[3]{9}}{2}$

(e) $\frac{\sqrt{x}}{7}$

10.(a) $a^2b^4c^5\sqrt[3]{a^2c}$

(b) $3x^4y^4\sqrt{3y^3}$

11.(a) $\frac{2\sqrt{7}}{7}$

(b) $\frac{2\sqrt{22xy}}{y}$

(c) $\frac{\sqrt[3]{45}}{3}$

(d) $\frac{\sqrt[3]{100xy}}{5y}$

12.(a) $-2\sqrt{10}$

(b) $53\sqrt{2}$

(c) $12a^2b\sqrt{5a}$

(d) $-5ab\sqrt[3]{2b}$

13.(a) $\frac{15}{2}\sqrt{2}$

(b) $\frac{7\sqrt[3]{9}}{12}$

(c) $(4\sqrt{x}-\sqrt{2x})\frac{1}{x}$

14.(a) $12\sqrt{35}$

(b) $5\sqrt{6}+8$

(c) $26-2\sqrt{15}$

(d) $14+6\sqrt{5}$

(e) $a-b$

15.(a) $\frac{12+2\sqrt{2}}{17}$

(b) $-9-4\sqrt{5}$

(c) $\frac{x\sqrt{x}+x\sqrt{y}}{x-y}$

16.(a) 5

(b) $\varnothing$

(c) 3

(d) 4

(e) 3

(f) 6

(g) −11

17.(a) $7i$

(b) $i\sqrt{6}$

18.(a) i

(b) 1

(c) −1

(d) $-i$

19.(a) $11-6i$

(b) $-4-15i$

(c) $8-4i$

(d) $4+7i$

20.(a) $30+10i$

(b) 13

21.(a) $-6i+3$

(b) $\frac{2+11i}{5}$

7

Quadratic Equations and Inequalities

7.1 SOLVING QUADRATIC EQUATIONS BY COMPLETING THE SQUARE

In Chapter 3 we solved quadratic equations by isolating 0 on one side of the equation, factoring, and setting each factor equal to 0. However, not all quadratic equations can be factored. Sections 7.1 and 7.2 discuss other ways to solve quadratic equations.

The square root property

One method used to solve quadratic equations makes use of the square root property.

The Square Root Property:
If $a^2 = b$, then $a = \sqrt{b}$ or $a = -\sqrt{b}$.

If $b \geq 0$, the solutions are real numbers. If $b < 0$, the solutions are complex numbers.

EXAMPLE 1 Solve.

a) $p^2 = 25$

b) $(2m - 6)^2 = 36$

c) $x^2 + 5 = 0$

d) $(3a + 4)^2 = 15$

SOLUTION 1

a) $p^2 = 25$

$p = \sqrt{25}$ or $p = -\sqrt{25}$ — Use the square root property.

$p = 5$ or $p = -5$ — Simplify each radical.

Check:

If $p = 5$ — If $p = -5$

$5^2 \stackrel{?}{=} 25$ — $(-5)^2 \stackrel{?}{=} 25$ — Substitute into the original equation.

$25 = 25$ — $25 = 25$

Both checks are true, so the solution set is $\{-5, 5\}$ which can also be written as $\{\pm 5\}$ where $\pm$ is read "plus or minus".

b) $(2m - 6)^2 = 36$

$2m - 6 = \sqrt{36}$ or $2m - 6 = -\sqrt{36}$ — Use the square root property.

$2m - 6 = 6$ or $2m - 6 = -6$ — Simplify each radical.

$2m = 12$ or $2m = 0$ — Add 6.

$m = 6$ or $m = 0$ — Divide by 2.

Check:

If $m = 6$, — If $m = 0$

$(2(6) - 6)^2 \stackrel{?}{=} 36$ — $(2(0) - 6)^2 \stackrel{?}{=} 36$

$(12 - 6)^2 \stackrel{?}{=} 36$ — $(0 - 6)^2 \stackrel{?}{=} 36$

$6^2 \stackrel{?}{=} 36$ — $(-6)^2 \stackrel{?}{=} 36$

$36 = 36$ — $36 = 36$

Both checks are true, so the solution set is $\{0, 6\}$.

c) $x^2 + 5 = 0$

$x^2 = -5$			Isolate the squared term, x^2, before applying the square root property.
$x = \sqrt{-5}$	or	$x = -\sqrt{-5}$	Use the square root property.
$x = i\sqrt{5}$	or	$x = -i\sqrt{5}$	Simplify each radical using $i = \sqrt{-1}$.

Both solutions check (look back at section 6.7 to recall what $(i\sqrt{5})^2$ will yield), so the solution set is $\{-i\sqrt{5}, i\sqrt{5}\}$ *or* $\{\pm i\sqrt{5}\}$.

d) $(3a+4)^2 = 15$

$3a + 4 = \sqrt{15}$	or	$3a + 4 = -\sqrt{15}$	Use the square root property.
$3a = -4 + \sqrt{15}$		$3a = -4 - \sqrt{15}$	Subtract 4.
$a = \dfrac{-4+\sqrt{15}}{3}$		$a = \dfrac{-4-\sqrt{15}}{3}$	Divide by 3.

Both solutions check, so the solution set is $\{\dfrac{-4-\sqrt{15}}{3}, \dfrac{-4+\sqrt{15}}{3}\}$ or $\{\dfrac{-4\pm\sqrt{15}}{3}\}$.

Completing the square

The equations in the previous example were quadratic equations, written in a way that allowed us to use the square root property. The technique called *completing the square* provides a method for writing every quadratic equation in that form.

To Solve a Quadratic Equation by Completing the Square:

1. Isolate the x^2 and x term on one side of the equation so the equation has the form $ax^2 + bx = c$.
2. If $a \neq 1$, divide every term by a, the coefficient of x^2. The equation should now be in the form $x^2 + \frac{b}{a}x = \frac{c}{a}$.
3. Add $(\frac{1}{2}$ times the coefficient of $x)^2$ to both sides of the equation.
4. Simplify the right side of the equation.
5. Write the left side as a perfect square trinomial.
6. Solve using the Square Root Property.
7. Check the answers in the original equation.

EXAMPLE 2 Solve by completing the square.

a) $m^2 + 5m - 3 = 0$

b) $2y^2 - 5y + 1 = 0$

c) $3x^2 + 5x + 3 = 0$

SOLUTION 2

a) $m^2 + 5m - 3 = 0$

$m^2 + 5m = 3$ — Isolate $m^2 + 5m$ by adding 3 to both sides.

$m^2 + 5m + \frac{25}{4} = 3 + \frac{25}{4}$ — $(\frac{1}{2} \cdot 5)^2 = (\frac{5}{2})^2 = \frac{25}{4}$.

$m^2 + 5m + \frac{25}{4} = \frac{37}{4}$ — $\frac{3}{1} + \frac{25}{4} = \frac{12}{4} + \frac{25}{4} = \frac{37}{4}$.

$(m + \frac{5}{2})^2 = \frac{37}{4}$ — Factor the left side.

$m + \frac{5}{2} = \sqrt{\frac{37}{4}}$ or $m + \frac{5}{2} = -\sqrt{\frac{37}{4}}$ — Use the square root property.

$m + \frac{5}{2} = \frac{\sqrt{37}}{2}$ $\quad m + \frac{5}{2} = \frac{-\sqrt{37}}{2}$ — $\sqrt{\frac{37}{4}} = \frac{\sqrt{37}}{\sqrt{4}} = \frac{\sqrt{37}}{2}$.

$m = -\frac{5}{2} + \frac{\sqrt{37}}{2}$ $\quad m = -\frac{5}{2} - \frac{\sqrt{37}}{2}$ $\quad$ Subtract $\frac{5}{2}$.

$m = \frac{-5+\sqrt{37}}{2}$ $\quad m = \frac{-5-\sqrt{37}}{2}$

The solution set is $\{m = \frac{-5+\sqrt{37}}{2}, m = \frac{-5-\sqrt{37}}{2}\}$ or $\{m = \frac{-5 \pm \sqrt{37}}{2}\}$.

b) $2y^2 - 5y + 1 = 0$

$2y^2 - 5y = -1$ $\quad$ Subtract 1 from both sides.

$\frac{2y^2}{2} - \frac{5y}{2} = -\frac{1}{2}$ $\quad$ Divide each term by 2.

$y^2 - \frac{5}{2}y = -\frac{1}{2}$

$y^2 - \frac{5}{2}y + \frac{25}{16} = -\frac{1}{2} + \frac{25}{16}$ $\quad$ $\left(\frac{1}{2} \cdot \frac{-5}{2}\right)^2 = \left(\frac{-5}{4}\right)^2 = \frac{25}{16}$

$y^2 - \frac{5}{2}y + \frac{9}{16} = \frac{17}{16}$ $\quad$ $-\frac{1}{2} + \frac{25}{16} = -\frac{8}{16} + \frac{25}{16} = \frac{17}{16}$.

$\left(y - \frac{3}{4}\right)^2 = \frac{17}{16}$ $\quad$ Factor the left side.

$y - \frac{3}{4} = \sqrt{\frac{17}{16}}$ or $\quad y - \frac{3}{4} = -\sqrt{\frac{17}{16}}$ $\quad$ Use the square root property.

$y - \frac{3}{4} = \frac{\sqrt{17}}{4}$ $\quad y - \frac{3}{4} = \frac{-\sqrt{17}}{4}$ $\quad$ $\sqrt{\frac{17}{16}} = \frac{\sqrt{17}}{\sqrt{16}} = \frac{\sqrt{17}}{4}$.

$y = \frac{3}{4} + \frac{\sqrt{17}}{4}$ $\quad y = \frac{3}{4} - \frac{\sqrt{17}}{4}$ $\quad$ Add $\frac{3}{4}$.

$y = \frac{3+\sqrt{17}}{4}$ or $\quad y = \frac{3-\sqrt{17}}{4}$

The solution set is $\{y = \frac{3+\sqrt{17}}{4}, y = \frac{3-\sqrt{17}}{4}\}$ or $\{y = \frac{3 \pm \sqrt{17}}{4}\}$.

c) $3x^2 + 5x + 3 = 0$

$$3x^2 + 5x = -3$$ Subtract 3 from both sides.

$$\frac{3x^2}{3} + \frac{5x}{3} = -\frac{3}{3}$$ Divide each term by 3.

$$x^2 + \frac{5}{3}x = -1$$

$$x^2 + \frac{5}{3}x + \frac{25}{36} = -1 + \frac{5}{36}$$ $$\left(\frac{1}{2} \cdot \frac{5}{3}\right)^2 = \left(\frac{5}{6}\right)^2 = \frac{25}{36}.$$

$$x^2 + \frac{5}{3}x + \frac{25}{6} = -\frac{11}{36}$$ $$-1 + \frac{25}{36} = -\frac{36}{36} + \frac{25}{36} = -\frac{11}{36}.$$

$$\left(x + \frac{5}{6}\right)^2 = -\frac{11}{36}$$ Factor the left side.

$$x + \frac{5}{6} = \sqrt{-\frac{11}{36}} \text{ or } x + \frac{5}{6} = -\sqrt{-\frac{11}{36}}$$ Use the square root property.

$$x + \frac{5}{6} = \frac{i\sqrt{11}}{6} \quad x + \frac{5}{6} = -\frac{i\sqrt{11}}{6}$$ $$\sqrt{-\frac{11}{36}} = \frac{\sqrt{-11}}{\sqrt{36}} = \frac{i\sqrt{11}}{6}.$$

$$x = -\frac{5}{6} + \frac{i\sqrt{11}}{6} \quad x = -\frac{5}{6} - \frac{i\sqrt{11}}{6}$$ Subtract $\frac{5}{6}$.

$$x = \frac{-5 + i\sqrt{11}}{6} \text{ or } x = \frac{-5 - i\sqrt{11}}{6}$$

The solution set is $\{x = \frac{-5 + i\sqrt{11}}{6}, x = \frac{-5 - i\sqrt{11}}{6}\}$ or $\{\frac{-5 \pm i\sqrt{11}}{6}\}$.

7.2 SOLVING QUADRATIC EQUATIONS USING THE QUADRATIC FORMULA

Although completing the square can be used to solve any quadratic equation, it can be a long process. The quadratic formula can also be used to solve any quadratic equation, but you MUST memorize the formula to use it!

The Quadratic Formula

The quadratic formula is derived by completing the square on $ax^2 + bx + c = 0$.

The Quadratic Formula.
The solutions to $ax^2 + bx + c = 0, a \neq 0$ are

$$x = \frac{-b \pm \sqrt{b^2 - 4ac}}{2a}$$

Solving quadratic equations using the quadratic formula

To Solve a Quadratic Equation Using the Formula

1. Write the equation in standard form: $ax^2 + bx + c = 0$. Multiply both sides by the LCD to clear fractions, if necessary.
2. Write down the values for a, b and c.
3. Substitute the values for a, b, and c into the Quadratic Formula.
4. Simplify the radicand.
5. Simplify the radical, if possible.
6. Factor the numerator to reduce, if possible.
7. Check the answers.

EXAMPLE 3 Solve each equation using the quadratic formula.

a) $2x^2 - 5x = 3$

b) $(2p - 1)(p + 3) = 4p - 2$

c) $\frac{1}{5}s^2 + s = -\frac{7}{20}$

d) $\frac{2x}{x-2} - \frac{6}{x-3} = -\frac{9}{(x-2)(x-3)}$

SOLUTION 3

a) $2x^2 - 5x = 3$

$2x^2 - 5x - 3 = 0$	Write the equation in standard form by subtracting 3 from both sides.
$a = 2, b = -5, c = -3$	Identify the values for a, b, and c.
$x = \dfrac{-b \pm \sqrt{b^2 - 4ac}}{2a}$	Write the quadratic formula.
$x = \dfrac{-(-5) \pm \sqrt{(-5)^2 - 4(2)(-3)}}{2(2)}$	Substitute the values for a, b, and c.
$x = \dfrac{5 \pm \sqrt{25 + 24}}{4}$	Simplify the radicand.
$x = \dfrac{5 \pm \sqrt{49}}{4}$	Simplify the radicand.
$x = \dfrac{5 \pm 7}{4}$	Simplify the radical.
$x = \dfrac{5 + 7}{4}$ or $x = \dfrac{5 - 7}{4}$	Solve for x.
$x = \dfrac{12}{4} = 3$ or $x = \dfrac{-2}{4} = \dfrac{-1}{2}$	

Check:

If $x = 3$	If $x = -\frac{1}{2}$
$2(3)^2 - 5(3) \stackrel{?}{=} 3$	$2(-\frac{1}{2})^2 - 5(-\frac{1}{2}) \stackrel{?}{=} 3$
$2(9) - 5(3) \stackrel{?}{=} 3$	$2(\frac{1}{4}) - 5(-\frac{1}{2}) \stackrel{?}{=} 3$
$18 - 15 \stackrel{?}{=} 3$	$\frac{1}{2} + \frac{5}{2} \stackrel{?}{=} 3$
$3 = 3$	$3 = 3$

Both solutions check, so the solution set is $\{-\frac{1}{2}, 3\}$.

b) $(2p - 1)(p + 3) = 4p - 2$

$2p^2 + 6p - p - 3 = 4p - 2$	To write the equation in standard form, multiply the left side.
$2p^2 + 5p - 3 = 4p - 2$	Combine similar terms.
$2p^2 + p - 1 = 0$	The equation is now in standard form.
$a = 2, b = 1, c = -1$	Identify *a*, *b*, and *c*.
$p = \dfrac{-b \pm \sqrt{b^2 - 4ac}}{2a}$	Write the quadratic formula. We are solving for *p*.
$p = \dfrac{-1 \pm \sqrt{1^2 - 4(2)(-1)}}{2(2)}$	Substitute the values for *a*, *b*, and *c*.
$p = \dfrac{-1 \pm \sqrt{1+8}}{4}$	Simplify the radicand.
$p = \dfrac{-1 \pm \sqrt{9}}{4}$	Simplify the radicand.
$p = \dfrac{-1 \pm 3}{4}$	Simplify the radical.
$p = \dfrac{-1+3}{4}$ or $p = \dfrac{-1-3}{4}$	Solve for *p*.
$p = \dfrac{2}{4} = \dfrac{1}{2}$ or $p = -\dfrac{4}{4} = -1$	

Try the check on your own. The solution set is $\{-1, \frac{1}{2}\}$.

c) $\frac{1}{5}s^2 + s = -\frac{7}{20}$

$20(\frac{1}{5}s^2 + 2) = 20(-\frac{7}{20})$	Multiply both sides by the LCD.
$4s^2 + 20s = -7$	Add 7 to both sides.
$4s^2 + 20s + 7 = 0$	The equation is now in standard form.
$a = 4, b = 20, c = 7$	Identify *a*, *b*, and *c*.

$$s = \frac{-b \pm \sqrt{b^2 - 4ac}}{2a}$$

Write the quadratic formula.

$$s = \frac{-20 \pm \sqrt{(20)^2 - 4(4)(7)}}{2(4)}$$

Substitute the values for a, b, and c.

$$s = \frac{-20 \pm \sqrt{400 - 112}}{8}$$

Simplify the radicand.

$$s = \frac{-20 \pm \sqrt{288}}{8}$$

Simplify the radicand.

$$s = \frac{-20 \pm 12\sqrt{2}}{8}$$

$$\sqrt{288} = \sqrt{144 \cdot 2} = \sqrt{144}\sqrt{2} = 12\sqrt{2}.$$

$$s = \frac{4(-5 \pm 3\sqrt{2})}{8}$$

Factor the numerator.

$$s = \frac{-5 \pm 3\sqrt{2}}{2}$$

Reduce $\frac{4}{8} = \frac{1}{2}$.

d) $\frac{2x}{x-2} - \frac{6}{x-3} = -\frac{9}{(x-2)(x-3)}$

$$(x-2)(x-3)\left(\frac{2x}{x-2} - \frac{6}{x-3}\right) = (x-2)(x-3)\left(-\frac{9}{(x-2)(x-3)}\right)$$

Multiply both sides by the LCD.

$$2x(x-3) - 6(x-2) = -9$$

$$2x^2 - 6x - 6x + 12 = -9$$

Simplify the left side.

$$2x^2 - 12x + 12 = -9$$

Combine similar terms.

$$2x^2 - 12x + 21 = 0$$

The equation is now in standard form.

$a = 2, b = -12, c = 21$

Identify a, b, and c.

$$x = \frac{-b \pm \sqrt{b^2 - 4ac}}{2a}$$

Write the quadratic formula.

$$x = \frac{-(-12) \pm \sqrt{(-12)^2 - 4(2)(21)}}{2(2)}$$

Substitute the values for a, b, and c.

$$x = \frac{12 \pm \sqrt{144 - 168}}{4}$$ Simplify the radicand.

$$x = \frac{12 \pm \sqrt{-24}}{4}$$ Simplify the radicand.

$$x = \frac{12 \pm 2i\sqrt{6}}{4}$$ $\sqrt{-24} = \sqrt{(-1) \cdot 24}.$

$$= i\sqrt{24} = 2i\sqrt{6}$$

$$x = \frac{2(6 + i\sqrt{6})}{4}$$ Factor the numerator.

$$x = \frac{6 \pm i\sqrt{6}}{2}$$ Reduce $\frac{2}{4} = \frac{1}{2}$.

Solving word problems with the quadratic formula

Word problems often produce quadratic equations that can be solved using the quadratic formula. Be careful to check whether solutions are reasonable.

EXAMPLE 4 If an object is thrown downward with an initial velocity of 8 feet per second, the distance s that it travels in time t is given by the equation $s = 16t^2 + 8t$. How long does it take the object to fall 44 feet?

SOLUTION 4

Let $s = 44$ feet — $s = 16t^2 + 8t$

$44 = 16t^2 + 8t$

$16t^2 + 8t - 44 = 0$ — Write the equation in standard form.

$a = 16, b = 8, c = -44$ — Identify a, b, and c.

$$t = \frac{-b \pm \sqrt{b^2 - 4ac}}{2a}$$ Write the quadratic formula.

$$t = \frac{-8 \pm \sqrt{8^2 - 4(16)(-44)}}{2(16)}$$ Substitute the values for a, b, and c.

$$t = \frac{-8 \pm \sqrt{2880}}{32}$$ Simplify the radicand.

$$t = \frac{-8 \pm 24\sqrt{5}}{32}$$ Simplify the radical.

$$t = \frac{8(-1 \pm 3\sqrt{5})}{32}$$ Factor the numerator.

$$t = \frac{-1 \pm 3\sqrt{5}}{4}$$ Reduce $\frac{8}{32} = \frac{1}{4}$.

Time cannot be negative, so we eliminate the solution $t = \frac{-1 - 3\sqrt{5}}{4}$.
Since $\sqrt{5} \approx 2.23$,

$$t \approx \frac{-1 + 3(2.323)}{4} \approx 1.42$$

Therefore it takes the object approximately 1.42 seconds to reach the ground.

EXAMPLE 5 Joan and Ron are preparing a chemistry project. If they work together, it will take 3 hours to complete the project. If each worked alone, Ron could complete the project in 1 hour more than Joan. How long would it take Joan to complete the project alone?

SOLUTION 5

Let x represent the number of hours for Joan to complete the project alone. Then

$x + 1$ represents the number of hours for Ron to complete the project alone,

$\frac{1}{x}$ represents the amount of work completed by Joan in 1 hour,

$\frac{1}{x+1}$ represents the amount of work completed by Ron in 1 hour, and

$\frac{1}{x} + \frac{1}{x+1}$ represents the amount of work completed together in 1 hour.

Then,

$\frac{1}{x}+\frac{1}{x+1}=\frac{1}{3}$	In 1 hour they complete $\frac{1}{3}$ of the total project working together.
$3x(x+1)\left(\frac{1}{x}+\frac{1}{x+1}\right)=3x(x+1)\left(\frac{1}{3}\right)$	Multiply both sides by the LCD.
$3(x+1)(1)+3x(1)=x(x+1)$	Use the distributive property.
$3x+3+3x=x^2+x$	Simplify each side.
$6x+3=x^2+x$	Combine like terms.
$6x+3-6x-3=x^2+x-6x-3$	Get 0 on one side.
$0=x^2-5x-3$	Equation is now in standard form.
$a=1, b=-5, c=-3$	Identify a, b, and c.
$x=\frac{-b\pm\sqrt{b^2-4ac}}{2a}$	Write the quadratic formula.
$x=\frac{-(-5)\pm\sqrt{(-5)^2-4(1)(-3)}}{2(1)}$	Substitute a, b, and c.
$x=\frac{5\pm\sqrt{37}}{2}$	Simplify the radicand.

Since $\sqrt{37}\approx 6.08$, we must eliminate the solution

$$\frac{5-\sqrt{37}}{2}\approx\frac{5-6.08}{2}\approx -0.54$$

since the number of hours cannot be negative. So,

$$\frac{5+\sqrt{37}}{2}\approx\frac{5+6.08}{2}\approx 5.54$$

It would take Joan approximately 5.33 hours to complete the project working alone.

7.3 THE DISCRIMINANT AND SOLUTIONS OF QUADRATIC EQUATIONS

The discriminant

The radicand of the quadratic formula, $b^2 - 4ac$, is called the **discriminant**. We can use the discriminant to determine whether the solutions to a quadratic equation are rational or irrational or imaginary. The chart below summarizes the relationship between the discriminant and the solutions of a quadratic equation $ax^2 + bx + c = 0$ where a, b, and c are integers and $a \neq 0$.

Discriminant	Solutions
$b^2 - 4ac > 0$ and a perfect square	Two rational solutions
$b^2 - 4ac > 0$ and not a perfect square	Two irrational solutions
$b^2 - 4ac = 0$	One rational solution
$b^2 - 4ac < 0$	Two complex solutions

EXAMPLE 6 Use the discriminant to determine the number and type of solutions for each equation.

a) $2x^2 - 5x - 3 = 0$

b) $\frac{1}{5}s^2 + s = -\frac{7}{20}$

c) $4x^2 - 12x + 9 = 0$

d) $2x^2 - 12x = -21$

SOLUTION 6

a) $2x^2 - 5x - 3 = 0$

$a = 2, b = -5, c = -3$	Identify a, b, and c.
$b^2 - 4ac = (-5)^2 - 4(2)(-3)$	Find the discriminant.
$= 25 + 24$	Simplify.
$= 49$	

Since $49 > 0$ and is a perfect square, the equation willhave two rational solutions.

b) $\frac{1}{5}s^2 + s = -\frac{7}{20}$	Write the equation in standard form.
$20\left(\frac{1}{5}s^2 + s\right) = 20\left(-\frac{7}{20}\right)$	Multiply by the LCD.
$4s^2 + 20s = -7$	
$4s^2 + 7 = 0$	Equation is now in standard form.
$a = 4, b = 20, c = 7$	Identify a, b, and c.
$b^2 - 4ac = 20^2 - 4(4)(7)$	Find the discriminant.
$= 400 - 112$	Simplify.
$= 288$	

Since $288 > 0$ and is not a perfect square, the equation will have two irrational solutions.

c) $4x^2 - 12x + 9 = 0$

$a = 4, b = -12, c = 9$	Identify a, b, and c.
$b^2 - 4ac = (-12)^2 - 4(4)(9)$	Find the discriminant.
$= 144 - 144$	Simplify.
$= 0$	

Since the discriminant is 0, the equation will have one rational solution.

d) $2x^2 - 12x = -21$	Write the equation in standard form.
$2x^2 - 12x + 21 = 0$	
$a = 2, b = -12, c = 21$	Identify a, b, and c.
$b^2 - 4ac = (-12)^2 - 4(2)(21)$	Find the discriminant.
$= 144 - 168$	Simplify.
$= -24$	

Since $-24 < 0$, the equation will have two imaginary solutions.

You may want to compare your answers to parts a, b, and d to your work in Example 3 parts a, c, and d.

Finding equations using the discriminant

The discriminant can also be used to find a quadratic equation with a given number and type of solution(s) as demonstrated in the next example.

EXAMPLE 7 Find k so that $4x^2 + kx + 1 = 0$ has exactly one rational solution.

SOLUTION 7

A quadratic equation has exactly one rational solution when $b^2 - 4ac = 0$. We find the discriminant for $4x^2 + kx + 1 = 0$ and set it equal to 0.

$4x^2 + kx + 1 = 0$ — The equation is in standard form.

$a = 4, b = k, c = 1$ — Identify a, b, and c.

$b^2 - 4ac = k^2 - 4(4)(1)$ — Find the discriminant.

$= k^2 - 16$

$k^2 - 16 = 0$ — Set the discriminant equal to 0 and solve.

$k^2 = 16$

$k = 4$ or $k = -4$ — Use the square root property.

When k is 4 or −4 the equation will have exactly one rational solution.

Finding equations using solutions

The previous example demonstrated how the discriminant can be used to find a quadratic equation. We can also find a quadratic equation if we know the solutions to the equation.

EXAMPLE 8 Find a quadratic equation with solutions $x = 2$ and $x = -4$.

SOLUTION 8

We will reverse the process we used to factor and solve a quadratic equation.

If

$$x = 2 \qquad \text{or} \qquad x = -4$$
$$x - 2 = 0 \qquad\qquad x + 4 = 0$$

If $x - 2 = 0$ or $x + 4 = 0$, their product will equal 0, so

$(x-2)(x+4) = 0$	Zero-factor property.
$x^2 + 4x - 2x - 8 = 0$	Multiply.
$x^2 + 2x - 8 = 0$	Combine similar terms.

Note that this answer is not unique, which means there are other quadratic equations with solutions of 2 and -4. Any constant multiple of $x^2 + 2x - 8 = 0$ such as $2x^2 + 4x - 16 = 0$ and $-3x^2 - 6x + 24 = 0$ will also have solutions of 2 and -4.

EXAMPLE 9 Find a quadratic equation with solutions $x = \frac{2\sqrt{3}}{3}$ and $x = -\sqrt{3}$.

SOLUTION 9

$x = \frac{2\sqrt{3}}{3}$ or $x = -\sqrt{3}$	Given solutions.
$x - \frac{2\sqrt{3}}{3} = 0 \quad x + \sqrt{3} = 0$	Get 0 on one side.
$(x - \frac{2\sqrt{3}}{3})(x + \sqrt{3}) = 0$	Zero factor property.
$x^2 + \sqrt{3}x - \frac{2\sqrt{3}}{3}x - (\frac{2\sqrt{3}}{3} \cdot 3) = 0$	Use FOIL to multiply.
$x^2 + \sqrt{3}x - \frac{2\sqrt{3}}{3}x - 2 = 0$	$\frac{2\sqrt{3}}{3} \cdot \sqrt{3} = \frac{2 \cdot 3}{3} = 2.$
$3(x^2 + \sqrt{3}x - \frac{2\sqrt{3}}{3}x - 2) = 3(0)$	Multiply by the LCD.
$3x^2 + 3\sqrt{3}x - 2\sqrt{3}x - 6 = 0$	
$3x^2 + \sqrt{3}x - 6 = 0$	Combine similar terms.

EXAMPLE 10 Find a quadratic equation with solution set $\{2i, -2i\}$.

SOLUTION 10

Since $x = 2i$ or $x = -2i$	Given solutions.
$x - 2i = 0 \quad x + 2i = 0$	Get 0 alone on one side.

$(x-2i)(x+2i)=0$	Zero factor property.
$x^2+2xi-2xi-4i^2=0$	Use FOIL to multiply.
$x^2-4i^2=0$	Combine similar terms.
$x^2-4(-1)=0$	Use $i^2=-1$.
$x^2+4=0$	

7.4 SOLVING EQUATIONS QUADRATIC IN FORM

Several types of equations that are *not* quadratic equations can be solved by factoring, completing the square, or the quadratic formula by making an appropriate substitution. In the following examples we will make a substitution that will allow us to write the equation as $au^2 + bu + c = 0$ where u will be an algebraic expression.

Replacing an expression in parentheses

If an equation contains an expression in parentheses that is squared and also to the first power, let u equal the expression in the parentheses.

EXAMPLE 11 Solve.

a) $(2x+3)^2+2(2x+3)-8=0$

b) $12(x-2)^2+5(x-2)-3=0$

SOLUTION 11

a) $(2x+3)^2+2(2x+3)-8=0$

Let $u=2x+3$	Let u equal the expression in parentheses.
Then	
$u^2+2u-8=0$	Substitute u for $2x+3$.
$(u+4)(u-2)=0$	Factor.

$u+4=0$ or $u-2=0$ — Set each factor equal to 0.

$u=-4$ or $u=2$ — Solve for u.

Now replace u with $2x+3$:

$2x+3=-4$ or $2x+3=2$

$2x=-7$ $\quad$ $2x=-1$ — Subtract 3.

$x=-\frac{7}{2}$ or $x=-\frac{1}{2}$ — Divide by 2.

The solution set is $\{-\frac{7}{2}, -\frac{1}{2}\}$.

b) $12(x-2)^2+5(x-2)-3=0$

Let $u=x-2$ — Let u equal the expression in parentheses.

$12u^2+5u-3=0$ — Substitute u for $x-2$.

$(3u-1)(4u+3)=0$ — Factor.

$3u-1=0$ or $4u+3=0$ — Set each factor equal to 0.

$3u=1$ $\quad$ $4u=-3$

$u=\frac{1}{3}$ $\quad$ $u=-\frac{3}{4}$

Now replace u with $x-2$:

$x-2=\frac{1}{3}$ or $x-2=-\frac{3}{4}$ — Solve for x.

$x=\frac{1}{3}+2$ or $x=-\frac{3}{4}+2$

$x=\frac{7}{3}$ $\quad$ $x=\frac{5}{4}$

The solution set is $\{\frac{7}{3}, \frac{5}{4}\}$.

Replacing fractional exponents

If an equation contains fractional exponents where one fractional exponent is twice the other fractional exponent, we can often make a substitution that will result in an equation that is quadratic in form.

EXAMPLE 12 Solve.

a) $2a^{2/3}+9a^{1/3}-5=0$

b) $x + 3\sqrt{x} - 28 = 0$

SOLUTION 12

a) $2a^{2/3} + 9a^{1/3} - 5 = 0$

Note that $\frac{2}{3}$ is $2\left(\frac{1}{3}\right)$.

Let $u = a^{1/3}$ Then $u^2 = a^{2/3}$.

$2u^2 + 9u - 5 = 0$	Substitute.
$(2u - 1)(u + 5) = 0$	Factor.
$2u - 1 = 0$ or $u + 5 = 0$	Set each factor equal to 0.
$2u = 1$ $\quad u = -5$	Solve for u.
$u = \frac{1}{2}$	

Now replace u with $a^{1/3}$.

$a^{1/3} = \frac{1}{2}$ or $a^{1/3} = -5$

$(a^{1/3})^3 = \left(\frac{1}{2}\right)^3$ or $(a^{1/3})^3 = (-5)^3$ — Raise each side to the third power.

$a = \frac{1}{8}$ $\qquad a = -125$

The solution set is $\{-125, \frac{1}{8}\}$.

b) $x + 3\sqrt{x} - 28 = 0$

$x + 3x^{1/2} - 28 = 0$ — Rewrite the equation using fraction exponents.

Note that $1 = 2\left(\frac{1}{2}\right)$.

Let $u = x^{1/2}$. Then $u^2 = x$.

$u^2 + 3u - 28 = 0$	Substitute.
$(u + 7)(u - 4) = 0$	Factor.
$u + 7 = 0$ or $u - 4 = 0$	Set each factor equal to 0.
$u = -7$ or $\quad u = 4$	Solve for u.

Now replace u with $x^{1/2}$.

$x^{1/2} = -7$ or $x^{1/2} = 4$ Solve for x.

$(x^{1/2})^2 = (-7)^2 \quad (x^{1/2})^2 = 4^2$ Square both sides.

$x = 49$ or $x = 16$

Squaring both sides of an equation can lead to extraneous roots. Let'scheck these proposed solutions.

Check:

If $x = 49$	If $x = 16$
$49 + 3\sqrt{49} - 28 \stackrel{?}{=} 0$	$6 + 3\sqrt{16} - 28 \stackrel{?}{=} 0$
$49 + 3(7) - 28 \stackrel{?}{=} 0$	$16 + 3(4) - 28 \stackrel{?}{=} 0$
$49 + 21 - 28 \stackrel{?}{=} 0$	$16 + 12 - 28 \stackrel{?}{=} 0$
$42 = 0$	$0 = 0$
False	True

Since $x = 49$ leads to a false statement, there is only one solution, $x = 16$.

Replacing other exponents

The previous examples replaced an expression raised to a fractional exponent with u. We can use the same idea for integer exponents when one exponent is twice as large as another exponent on the same variable or expression in the equation.

EXAMPLE 13 Solve.

a) $4y^4 = 11y^2 + 3$

b) $9a^4 - 25 = 0$

SOLUTION 13

a) $4y^4 = 11y^2 + 3$

$4y^4 - 11y^2 - 3 = 0$ Get 0 alone on one side.

Let $u = y^2$. Then $u^2 = y^4$.

$4u^2 - 11u - 3 = 0$	Substitute.
$(4u + 1)(u - 3) = 0$	Factor.
$4u + 1 = 0$ or $u - 3 = 0$	Set each factor equal to 0.
$4u = -1 \qquad u = 3$	Solve for u.
$u = -\frac{1}{4}$	

Now replace u with y^2.

$y^2 = -\frac{1}{4}$ or $y^2 = 3$	Use the square root property to solve for y.
$y = \sqrt{-\frac{1}{4}}$ or $y = -\sqrt{-\frac{1}{4}}$ $y = \sqrt{3}$ or $y = -\sqrt{3}$	
$y = \frac{1}{2}i$ or $y = -\frac{1}{2}i$	Simplify the radical.

Try the check on your own. The solution set is $\{-\frac{1}{2}i, \frac{1}{2}i -\sqrt{3}, \sqrt{3}\}$.

b) $9a^4 - 25 = 0$

Let $u = a^2$. Then $u^2 = a^4$.

$9u^2 - 25 = 0$	Substitute.
$9u^2 = 25$	
$u^2 = \frac{25}{9}$	Solve using the square root property.
$u = \sqrt{\frac{25}{9}} \qquad u = -\sqrt{\frac{25}{9}}$	Simplify the radical.
$u = \frac{5}{3} \qquad u = -\frac{5}{3}$	

Now replace u with a^2.

$a^2 = \frac{5}{3} \qquad a^2 = -\frac{5}{3}$	Substitute.
$a = \sqrt{\frac{5}{3}}$ or $a = -\sqrt{\frac{5}{3}}$ $a = \sqrt{-\frac{5}{3}}$ or $a = -\sqrt{-\frac{5}{3}}$	

Use the square root property.

$$a = \frac{\sqrt{5}}{\sqrt{3}} \cdot \frac{\sqrt{3}}{\sqrt{3}} \quad a = -\left(\frac{\sqrt{5}}{\sqrt{3}} \cdot \frac{\sqrt{3}}{\sqrt{3}}\right)$$

$$a = \left(i\frac{\sqrt{5}}{\sqrt{3}}\right) \cdot \frac{\sqrt{3}}{\sqrt{3}} \quad a = \left(-i\frac{\sqrt{5}}{\sqrt{3}}\right) \cdot \frac{\sqrt{3}}{\sqrt{3}}$$

Rationalize the denominator.

$$a = \frac{\sqrt{15}}{3} \quad a = -\frac{\sqrt{15}}{3} \quad a = i\frac{\sqrt{15}}{3} \quad a = -\frac{i\sqrt{15}}{3}$$

Try the check on your own. The solution set is $\{\pm\frac{\sqrt{15}}{3}, \pm\frac{i\sqrt{15}}{3}\}$.

7.5 SOLVING QUADRATIC INEQUALITIES

Quadratic inequalities

A quadratic inequality can be written as $ax^2 + bx + c\square 0$, where the box contains an inequality symbol, <, >, ≤ or ≥. The technique we will use to solve quadratic inequalities can also be used to solve polynomial inequalities of higher degree and inequalities involving rational expressions that contain variables in one or more denominators.

To Solve Quadratic Inequalities:

1. Isolate 0 on the right side of the inequality.
2. Factor the quadratic expression on the left side.
3. Find the split points by setting each factor equal to 0 and solving.
4. Draw a dotted vertical line at each split point on a number line.
5. Write the factors on the left edge of the number line.
6. Pick a number in each region created by the split points, substitute it into each factor, and record the resulting sign (+ or –) on the number line.
7. Use the sign rules for products to determine regions to be shaded for the solution. Use closed dots on split points to be included in the solution (usually ≥ or ≤). Use open dots on split points that are not included in the solution (usually < or >).

EXAMPLE 14 Solve $x^2 - x - 6 > 0$.

SOLUTION 14

$x^2 - x - 6 > 0$ — 0 is isolated.

$(x-3)(x+2) > 0$ — Factor the left side.

$x - 3 = 0$ or $x + 2 = 0$ — Find split points.

$x = 3 \qquad x = -2$

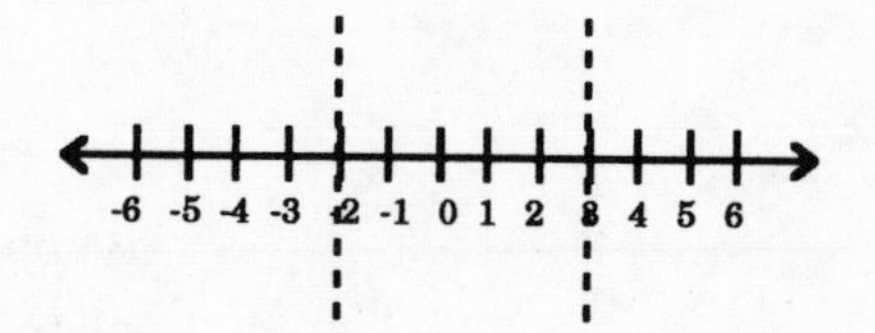

Draw dotted vertical lines at the split points. Write the factors on the left edge of the number line.

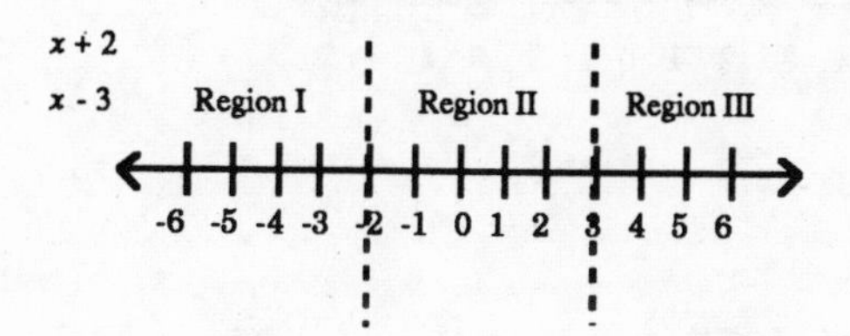

Pick a number in Region I, say –4, substitute into each factor and record the sign.

$x + 2 = -4 + 2 = -2$ Record

$-x - 3 = -4 - 3 = -7$

Record –

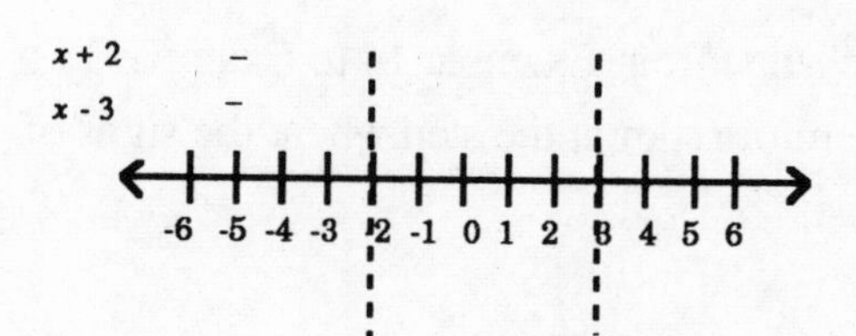

Pick a number in Region II, say 0, substitute into each factor and record the sign.

$x + 2 = 0 + 2 = +2$ Record +

$x - 3 = 0 - 3 = -3$ Record –

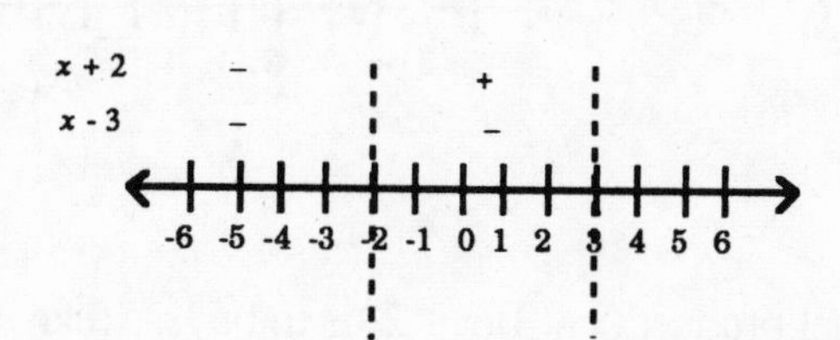

Pick a number in Region III, say 5, substitute into each factor and record the sign.

$x + 2 = 5 + 2 = +7$ Record +

$x - 3 = 5 - 3 = +2$ Record +

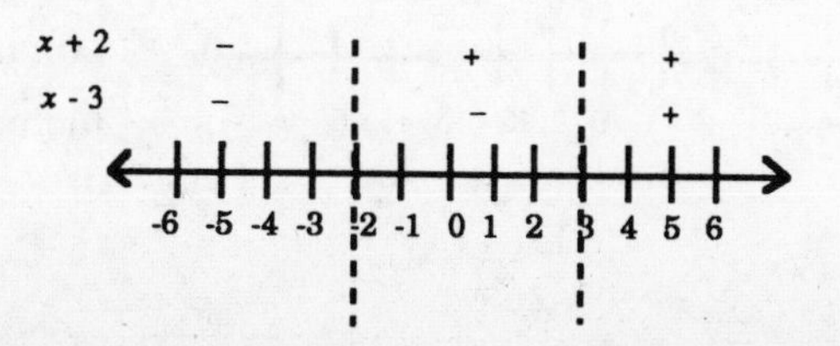

Use the sign rules for products to determine the sign of the product in each region.

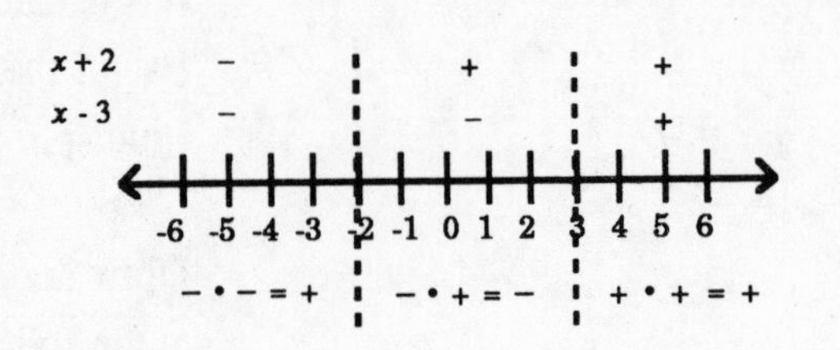

The original product $(x-3)(x+2)$ was to be greater than 0, so must be positive. Therefore, we shade regions I and III:

Use open dots on the split points because they are not included in the solution.

EXAMPLE 15 Solve $x^2 - x - 6 \leq 0$.

SOLUTION 15

The only change from Example 14 to Example 15 is the inequality symbol. We can therefore start at the step where the signs of the products are on the number line:

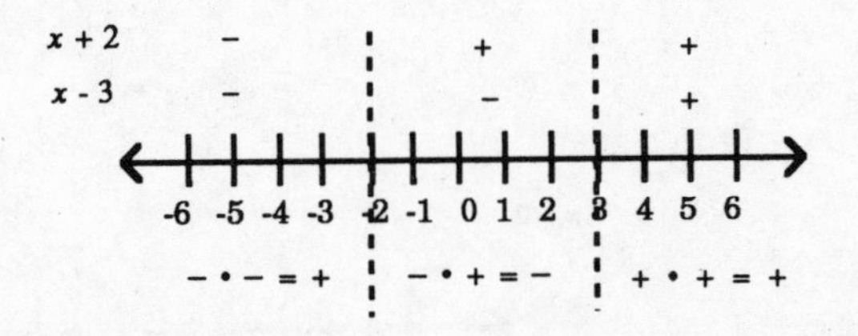

The original product $(x-3)(x+2)$ is to be less than or equal to 0, so must be negative. Therefore, we shade region II:

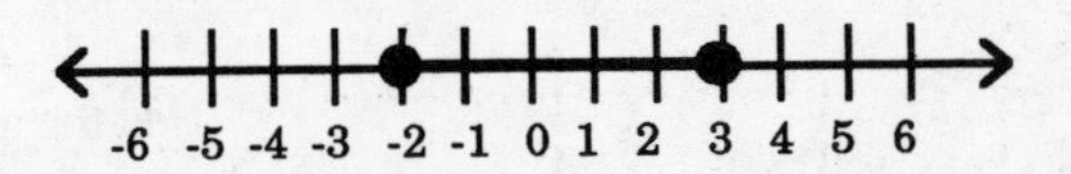

Use closed dots on the split points because they are included in the solution.

EXAMPLE 16 Solve $4x^2 - 7x \geq 2$.

SOLUTION 16

$4x^2 - 7x \geq 2$

$4x^2 - 7x - 2 \geq 0$ Isolate 0.

$(4x + 1)(x - 2) \geq 0$ Factor the left side.

$4x + 1 = 0$ $\quad$ $x - 2 = 0$ Find the split points.

$4x = -1$ $\quad$ $x = 2$

$x = -\frac{1}{4}$

Draw dotted vertical lines at the split points.

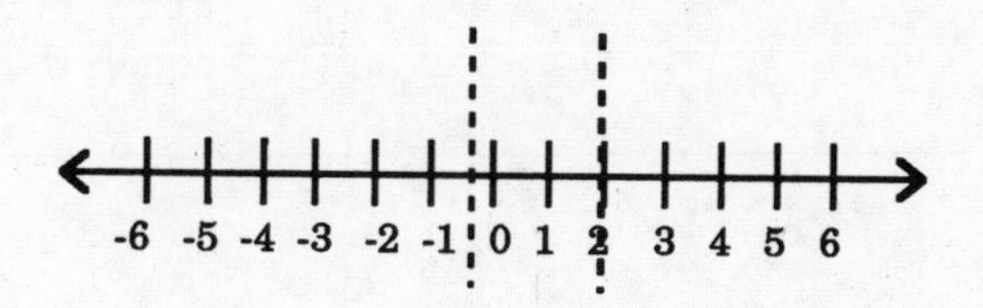

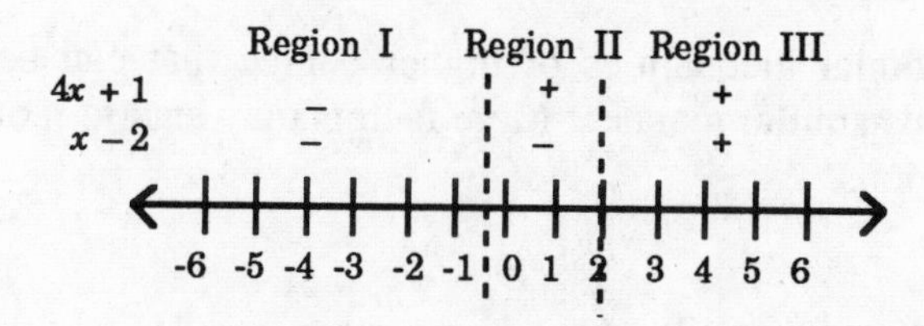

Region I: Let $x = -2$

$4x + 1 = 4(-2) + 1 = -7$ Record –.

$x - 2 = -2 - 2 = -4$ Record –.

Pick a number in each region.

Substitute that number into each factor and record the sign.

Region II: Let $x = 0$

$4x + 1 = 4(0) + 1 = +1$ Record +.

$x - 2 = 0 - 2 = -2$ Record –.

Region III: Let $x = 3$

$4(3) + 1 = 12 + 1 = +13$ Record +.

$x - 2 = 3 - 2 = +1$ Record +.

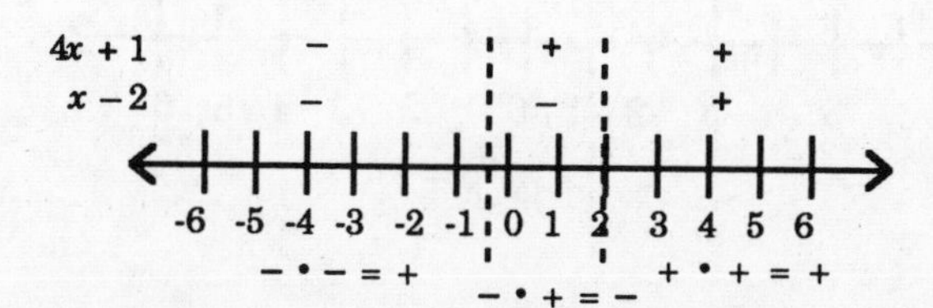

Use the sign rules for products to determine the sign of the product in each region.

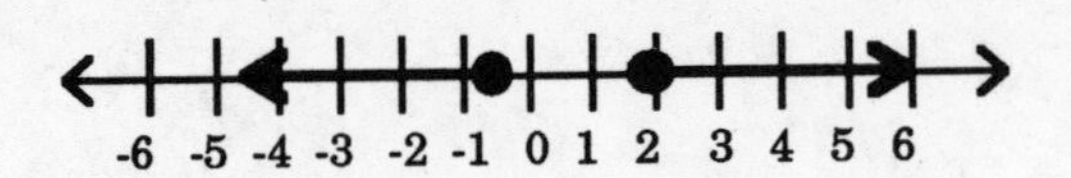

The original product $(4x + 1)(x - 2)$ was to be greater than or equal to 0, so must be positive or equal to 0. Therefore we shade Regions I and III.

Use closed dots on the split points because the product equals 0 there.

Higher degree inequalities

Polynomial inequalities of higher degree that can be factored can be solved in a similar manner. More factors may lead to more split points and more regions.

EXAMPLE 17 Solve.

a) $(x - 4)(x + 1)(x + 3) < 0$

b) $(x - 2)^2(2x + 1) > 0$

SOLUTION 17

a) $(x - 4)(x + 1)(x + 3) < 0$

$x - 4 = 0 \qquad x + 1 = 0 \qquad x + 3 = 0$

$x = 4 \qquad x = -1 \qquad x = -3$

Find the split points. Use vertical lines at each split point.

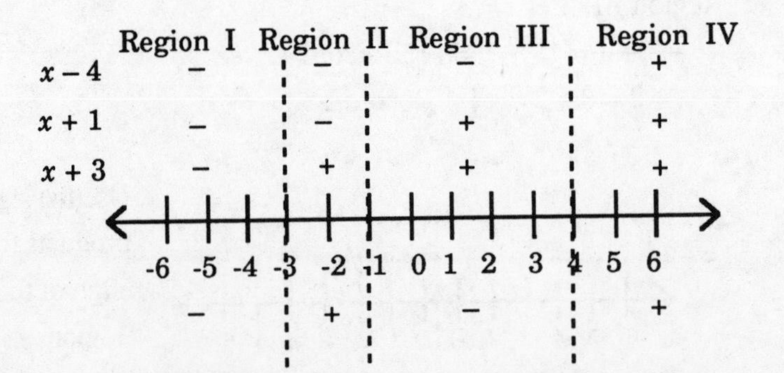

Region I: Let $x = -5$

$x - 4 = -5 - 4 = -9$ Record –.

$x + 1 = -5 + 1 = -4$ Record –. The product is –.

$x + 3 = -5 + 3 = -2$ Record –.

Region II: Let $x = -2$

$x - 4 = -2 - 4 = -6$ Record –.

$x + 1 = -2 + 1 = -1$ Record –. The product is +.

$x + 3 = -2 + 3 = +1$ Record +.

Region III: Let $x = 0$

$x - 4 = 0 - 4 = -4$ Record –.

$x + 1 = 0 + 1 = +1$ Record +. The product is –.

$x + 3 = 0 + 3 = +3$ Record +.

Region IV: Let $x = 6$

$x - 4 = 6 - 4 = +2$ Record +.

$x + 1 = 6 + 1 = +7$ Record +. The product is +.

$x + 3 = 6 + 3 = +9$ Record +.

The original product was to be less than 0, so must be negative. Therefore we shade Regions I and III:

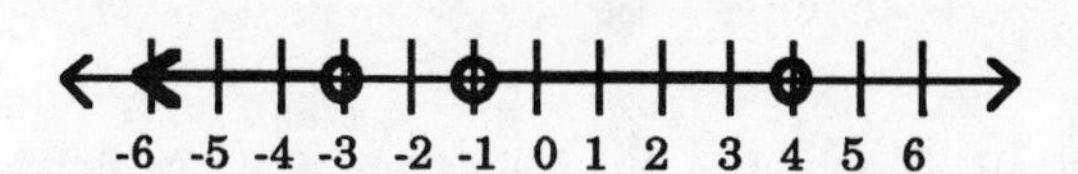

Use open dots on the split points because they are not included in the solution.

b) $(x - 2)^2(2x + 1) > 0$

$(x - 2)^2 = 0$ $\quad$ $2x + 1 = 0$ $\quad$ Find the split points.

$x - 2 = 0$ $\quad$ $2x = -1$

$x = 2$ $\quad$ $x = -\frac{1}{2}$

Use vertical lines at each split point.

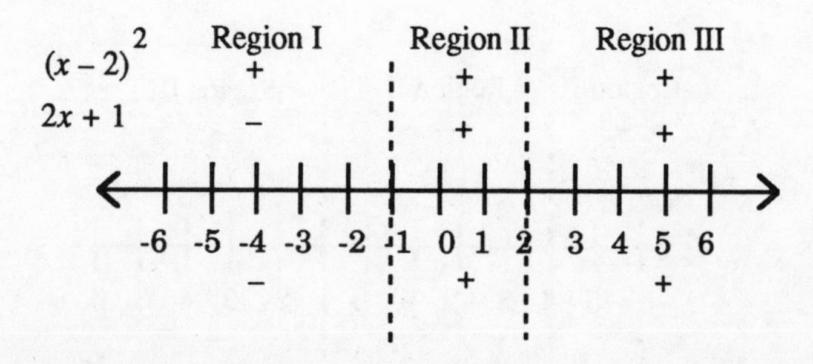

Region I: Let $x = -2$

$(x-2)^2 = (-2-2)^2 = (-4)^2 = +16$ Record +.

$2x + 1 = 2(-2) + 1 = -4 + 1 = -3$ Record –. The product is –.

Region II: Let $x = 0$

$(x-2)^2 = (0-2)^2 = (-2)^2 = +4$ Record +.

$2x + 1 = 2(0) + 1 = 0 + 1 = +1$ Record +. The product is +.

Region III: Let $x = 3$

$(x-2)^2 = (3-2)^2 = 1^2 = +1$ Record +.

$2x + 1 = 2(3) + 1 = 6 + 1 = +7$ Record +. The product is +.

The original product was to be greater than 0, so must be positive. Therefore we shade Regions II and III.

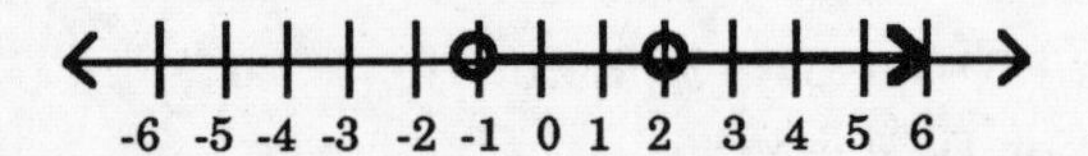

The split points have open dots. Note that there is an open dot at $x = 2$ since substituting $x = 2$ into the original inequality results in a false statement. Note also that the factor $(x-2)^2$ is always positive since the square of a real number is always positive.

Rational expressions

If an inequality contains a rational expression, we find split points by setting the numerator and denominator equal to 0 and solving. Then use the sign rules for quotients to determine regions to be shaded.

EXAMPLE 18 Solve $\frac{x-1}{x+4} > 0$.

SOLUTION 18

$x - 1 = 0$ $\quad$ $x + 4 = 0$ $\quad$ Find split points.

$x = 1$ $\quad$ $x = -4$

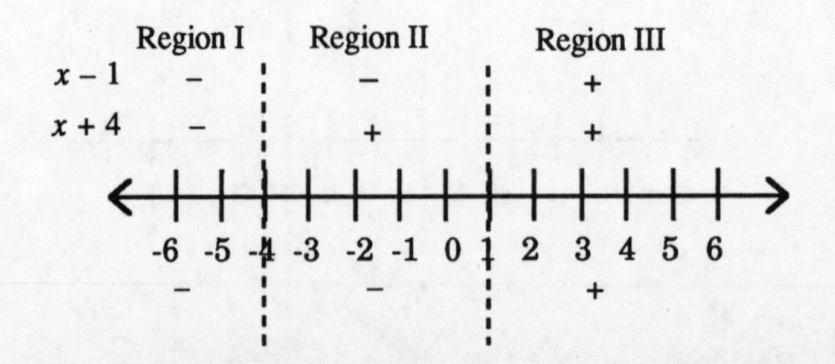

Region I: Let $x = -5$

$x - 1 = -5 - 1 = -6$ Record $-$.

$x + 4 = -5 + 4 = -1$ Record $-$. The quotient is $+$.

Region II: Let $x = 0$

$x - 1 = 0 - 1 = -1$ Record $-$.

$x + 4 = 0 + 4 = +4$ Record $+$. The quotient is $-$.

Region III: Let $x = 2$

$x - 1 = 2 - 1 = +1$ Record $+$.

$x + 4 = 2 + 4 = +6$ Record $+$. The quotient is $+$.

The quotient is to be greater than 0, so must be positive. Therefore we shade Regions I and III:

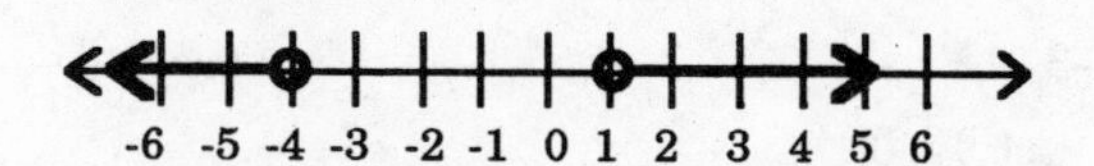

Use open dots on the split points because they are not included in the solution.

EXAMPLE 19 Solve $\dfrac{2}{x-1} \leq \dfrac{3}{x+2}$.

SOLUTION 19

$$\frac{2}{x-1} \leq \frac{3}{x+2}$$

$$\frac{2}{x-1} - \frac{3}{x+2} \leq 0$$ Isolate 0.

$$\frac{2(x+2)}{(x-1)(x+2)} - \frac{3(x-1)}{(x+2)(x-1)} \leq 0$$ Add the rational expressions using the LCD $(x-1)(x+2)$.

$$\frac{2x+4-3x+3}{(x-1)(x+2)} \leq 0$$

$$\frac{-x+7}{(x-1)(x+2)} \leq 0$$

$-x + 7 = 0$ $\quad x - 1 = 0$ $\quad x + 2 = 0$ Find the split points.

$x = 7$ $\quad x = 1$ $\quad x = -2$

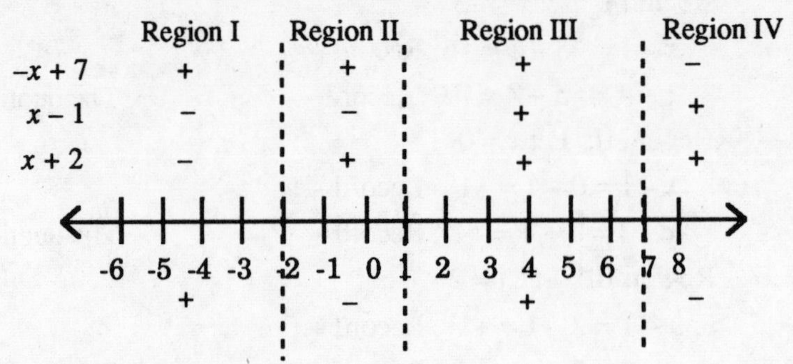

Region I: Let $x = -3$

$-x + 7 = -(-3) + 7 = +10$ Record +.

$x - 1 = -3 - 1 = -4$ Record –. The quotient is +.

$x + 2 = -3 + 2 = -1$ Record –.

Region II: Let $x = 0$

$-x + 7 = -0 + 7 = +7$ Record +.

$x - 1 = 0 - 1 = -1$ Record –. The quotient is –.

$x + 2 = 0 + 2 = +2$ Record +.

Region III: Let $x = 4$

$-x + 7 = -4 + 7 = +3$ Record +.

$x - 1 = 4 - 1 = +3$ Record +. The quotient is +.

$x + 2 = 4 + 2 = +6$ Record +.

Region IV: Let $x = 8$

$-x + 7 = -8 + 7 = -1$ Record –.

$x - 1 = 8 - 1 = +7$ Record +. The quotient is –.

$x + 2 = 8 + 2 = +10$ Record +.

The quotient needs to be less than or equal to 0, so must be negative or equal to 0. Therefore, Regions II and IV must be shaded:

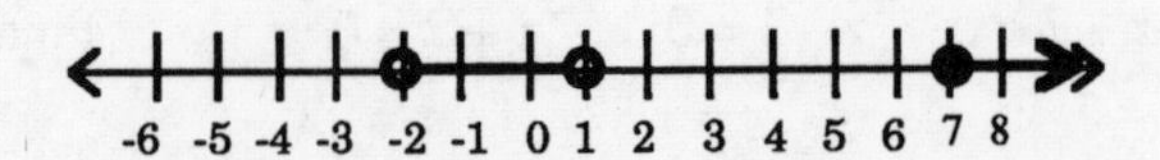

Note that we cannot use closed dots at $x = -2$ and $x = 1$ because they would make the original inequality undefined (we cannot divide by 0!).

Practice Exercises

1. Solve.

(a) $p^2 = 49$

(b) $(3m + 2)^2 = 144$

(c) $x^2 + 1 = 0$

(d) $(2a - 5)^2 = 21$

2. Solve by completing the square.

(a) $m^2 - 3m - 1 = 0$

(b) $4y^2 - 2y + 3 = 0$

(c) $16x^2 + 8x + 5 = 0$

3. Solve each equation using the quadratic formula.

(a) $6x^2 + 13x = 5$

(b) $(2p + 3)(p - 2) = 2p - 4$

(c) $\frac{1}{20}s^2 + \frac{1}{2}s = \frac{3}{10}$

(d) $\frac{4x}{x-1} - \frac{5}{x+2} = -\frac{6}{(x-1)(x+2)}$

4. Solve.

(a) If an object is thrown downward with an initial velocity of 6 feet per second, the distance s that it travels in time t is given by the equation $s = 16t^2 + 6t$. How long does it take the object to fall 10 feet?

(b) Maria and Lisa are preparing a slide presentation. If they work together, it will take 12 hours to complete the project. If each worked alone, Maria could complete then project in 3 hours more than Lisa. How long would it take Lisa to complete the project alone?

5. Use the discriminant to determine the number and type of solutions for each equation.

(a) $x^2 + 14x + 0 = 0$

(b) $\frac{1}{6}x^2 - \frac{1}{2}x = -\frac{13}{24}$

(c) $x^2 = 10x - 25$

(d) $8x^2 + 2x = 3$

6. Find k so that the given quadratic equation has the indicated number and type of solution(s).

(a) $9x^2 + kx + 16 = 0$, one rational solution

(b) $25x^2 - kx + 4 = 0$, one rational solution

7. Find a quadratic equation with the given solutions.

(a) $x = -1$ and $x = 3$

(b) $x = \frac{\sqrt{3}}{3}$ and $x = -2\sqrt{3}$

(c) $x = 3i$ and $x = -3i$

8. Solve.

(a) $(3x + 1)^2 - 9(3x + 1) + 20 = 0$

(b) $2(x - 5)^2 + 5(x - 5) - 3 = 0$

(c) $3a^{2/3} + 10a^{1/3} - 8 = 0$

(d) $x + 4\sqrt{x} - 12 = 0$

(e) $9y^4 + 44y^2 - 5 = 0$

(f) $16a^4 - 9 = 0$

9. Solve.

(a) $x^2 + 2x - 3 > 0$

(b) $3x^2 + 8x - 3 \le 0$

(c) $(x + 3)(x - 1)(x - 4) \ge 0$

(d) $(x + 2)^2(2x - 1) < 0$

Solutions

1.

(a) ± 7

(b) $-\frac{14}{3}, \frac{10}{3}$

(c) $\pm i$

(d) $\frac{5 \pm \sqrt{21}}{2}$

2.(a) $\frac{3 \pm \sqrt{13}}{2}$

(b) $\frac{3 \pm \sqrt{6}}{2}$

(c) $\frac{-1 \pm i}{4}$

3.(a) $-\frac{5}{2}, \frac{1}{3}$

(b) $-\frac{1}{2}, 2$

(c) $-5 \pm \sqrt{31}$

(d) $\frac{13 \pm i\sqrt{7}}{8}$

4.

(a) $\frac{5}{8}$ second

(b) approximately 22.6 hours

5.

(a) two irrational solutions

(b) two complex solutions

(c) one rational solution

(d) two rational solutions

6.

(a) 24

(b) 20

7.

(a) $x^2 - 2x - 3 = 0$

(b) $3x^2 + 2\sqrt{3}x - 6 = 0$

(c) $x^2 + 9 = 0$

8.

(a) $1, \frac{4}{3}$

(b) $2, \frac{11}{2}$

(c) $-64, \frac{8}{27}$

(d) 4

(e) $\pm\frac{1}{3}, \pm i\sqrt{5}$

(f) $\pm\frac{\sqrt{3}}{2}, \pm\frac{i\sqrt{3}}{2}$

9.

(a)

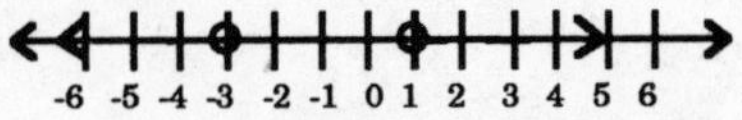

-6 -5 -4 -3 -2 -1 0 1 2 3 4 5 6

(b)

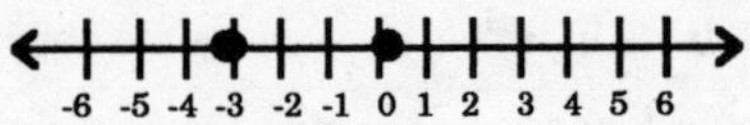

-6 -5 -4 -3 -2 -1 0 1 2 3 4 5 6

(c)

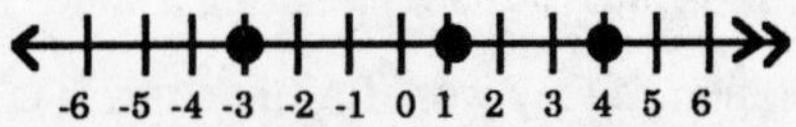

-6 -5 -4 -3 -2 -1 0 1 2 3 4 5 6

(d)

-6 -5 -4 -3 -2 -1 0 1 2 3 4 5 6

Solving systems of linear equations by addition

One technique used to solve systems of equations is called the addition method (or elimination method). The goal in this method is to add the equations together so that a variable is eliminated.

To Solve a System by Addition:

1. Write both equations in the form $ax + by = c$. Multiply the LCD to eliminate any fractions.
2. If coefficients of either x or y are additive inverses (like 4 and –4), add the equations. If coefficients are not additive inverses, eliminate x's by multiplying each equation by the coefficient of x from the other equation. If the signs are the same, multiply one equation by –1.
3. Add the equations together.
4. Solve for y.
5. Substitute the value for y in either of the original equations to solve for x.
6. Check your answer.

Note: You may eliminate y's in the same manner, solve for x, substitute, and solve for y. Depending on the coefficients of x and y, it is sometimes easier to eliminate y. Study the examples below, keeping in mind that each could be solved by eliminating the other variable.

EXAMPLE 2 Solve each system.

a) $3x + y = -1$

$x - y = -3$

b) $4x + 5y = 12$

$3x - 7y = -34$

c) $\frac{1}{4}x - \frac{1}{2}y = -\frac{5}{4}$

$\frac{1}{3}x + \frac{1}{4}y = -\frac{3}{4}$

d) $3x + 4y = 3$

$6x - 6y = -1$

SOLUTION 2

a) $3x + y = -1$ Because the coefficients of y

$x - y = -3$ — are additive inverses (1 and –1), add the equations.

$$\begin{aligned} 3x + y &= -1 \\ \underline{x - y} &\underline{= -3} \\ 4x \quad &= -4 \end{aligned}$$

Solve for x.

$x = -1$

$3x + y = -1$ — Use either original equation to solve for y.

$3(-1) + y = -1$ — Substitute $x = -1$.

$-3 + y = -1$ — Add 3 to both sides.

$y = 2$

$(-1, 2)$ — Proposed solution.

We solved and checked this as Example 1. Compare the techniques. Which do you prefer?

b) $4x + 5y = 12$ — Let's eliminate y.

$3x - 7y = -34$

$7(4x + 5y = 12) \Rightarrow 28x + 35y = 84$ — Multiply by the coefficients of y.

$5(3x - 7y = -34) \Rightarrow 15x - 35y = -170$

$$\begin{aligned} 28x + 35y &= 84 \\ \underline{15x - 35y} &\underline{= -170} \\ 43x \quad &= -86 \end{aligned}$$

Add the equations.

Solve for x.

$x = -2$

$4x + 5y = 12$ — Use one of the original equations to solve for y.

$4(-2) + 5y = 12$ — Substitute $x = -2$.

$-8 + 5y = 12$ — Add 8 to both sides.

$5y = 20$ — Divide by 5.

$y = 4$

$(-2, 4)$ — Proposed solution.

Check:

$3x - 7y = -34$ — Use the other original equation to check.

$3(-2) - 7(4) \stackrel{?}{=} -34$

$-6 - 28 \stackrel{?}{=} -34$

$-34 = -34$ True.

The solution to the system is (–2, 4).

c) $\frac{1}{4}x - \frac{1}{2}y = -\frac{5}{4}$

$\frac{1}{3}x + \frac{1}{4}y = -\frac{3}{4}$

Before eliminating a variable, clear fractions by multiplying each equation by its LCD.

$4\left(\frac{1}{4}x - \frac{1}{2}y\right) = 4\left(-\frac{5}{4}\right) \quad \Rightarrow \quad x - 2y = -5$

$12\left(\frac{1}{3}x + \frac{1}{4}y\right) = 12\left(-\frac{3}{4}\right) \quad \Rightarrow \quad 4x + 3y = -9$

$x - 2y = -5$ Let's eliminate x.

$4x + 3y = -9$

$4(x - 2y = -5) \quad \Rightarrow \quad 4x - 8y = -20$ Multiply by the coefficients

$1(4x + 3y = -9) \quad \Rightarrow \quad 4x + 3y = -9$ of x.

$-1(4x - 8y = -20) \Rightarrow -4x + 8y = 20$ Since the signs are the same, multiply one equation by –1.

$-4x + 8y = 20$ Add the equations.

$\underline{4x + 3y = -9}$

$11y = 11$ Solve for y.

$y = 1$

Rather than substitute $y = 1$ into one of the original equations to find x, it may be easier to eliminate y:

$3(x - 2y = -5) \quad \Rightarrow \quad 3x - 6y = -15$ Multiply by the coefficients

$2(4x + 3y = -9) \Rightarrow \quad 8x + 6y = -18$ of y.

$3x - 6y = -15$ Add the equations.

$8x + 6y = -18$

$11x \quad = -33$ Solve for x.

$x = -3$

(–3, 1) Proposed solution.

Try the check on your own. The solution to the system is (–3, 1).

d) $3x + 4y = 3$ — Let's eliminate y.

$6x - 6y = -1$

$6(3x + 4y = 3) \Rightarrow 18x + 24y = 18$ — Multiply by the coefficients of y.

$4(6x - 6y = -1) \Rightarrow 24x - 24y = -4$

$$\begin{array}{l} 18x + 24y = 18 \\ \underline{24x - 24y = -4} \\ 42x \qquad\quad = 14 \end{array}$$

Add the equations.

Solve for x.

$$x = \frac{14}{42} = \frac{1}{3}$$

$3x + 4y = 3$ — Use either original equation to solve for y.

$3\left(\frac{1}{3}\right) + 4y = 3$ — Substitute $x = \frac{1}{3}$.

$1 + 4y = 3$ — Subtract 1 from both sides.

$4y = 2$ — Divide both sides by 4.

$y = \frac{1}{2}$

Check:

$6x - 6y = -1$ — Use the other original equation to check.

$$6\left(\frac{1}{3}\right) - 6\left(\frac{1}{2}\right) \stackrel{?}{=} -1$$

$$2 - 3 \stackrel{?}{=} -1$$

$-1 = -1$ — True.

The solution to the system is $\left(\frac{1}{3}, \frac{1}{2}\right)$.

Determining when a system is inconsistent or dependent

The graph of the solutions to the systems of parallel lines shows there were no solutions; similarly, the graph of a dependent system (coinciding lines) showed there were an infinite number of solutions. The following table will help you decide when a system is consistent and independent, inconsistent, or dependent when using the addition method.

Look For	Graph	Solution
Only one variable eliminated	Consistent and independent -- two intersecting lines	One ordered-pair solution
Both variables eliminated and resulting statement is true	Dependent equations -- lines coincide	Infinite number of solutions
Both variables eliminated and resulting statement is false	Inconsistent system -- parallel lines	No solution

EXAMPLE 3 Solve each system.

a) $3x + y = 4$

$6x + 2y = 8$

b) $5x - y = 8$

$10x - 2y = 4$

SOLUTION 3

a) $3x + y = 4$ — Let's eliminate y.

$6x + 2y = 8$

$-2(3x + y = 4) \Rightarrow -6x - 2y = -8$ — Since the signs are the same,

$1(6x + 2y = 8) \Rightarrow 6x + 2y = 8$ — combine multiplying by the coefficients and multiplying one equation by -1 into one step.

$$\begin{aligned} -6x - 2y &= -8 \\ \underline{6x + 2y} &= \underline{8} \\ 0 &= 0 \end{aligned}$$

Add the equations.

Both variables were eliminated and the resulting statement is true. This is a dependent system with an infinite number of solutions. The solution set can be written as $\{(x, y) \mid 3x + y = 4\}$ or $\{(x, y) \mid 6x + 2y = 8\}$.

b) $5x - y = 8$ — Let's eliminate y.

$10x - 2y = 4$

$-2(5x - y = 8) \Rightarrow -10x + 2y = -16$ — Multiply by the coefficients of y and multiply one equation by -1.

$1(10x - 2y = 4) \Rightarrow 10x - 2y = 4$

$$\begin{array}{r} -10x + 2y = -16 \\ \underline{10x - 2y = \quad 4} \\ 0 = -12 \end{array}$$

Add the equations.

Both variables were eliminated and the resulting statement is false. This is an inconsistent system with no solutions. The solution set is the empty set, $\varnothing$.

Solving systems by substitution

If one of the equations is solved for a variable or can be easily solved for a variable, the substitution method may be more convenient to use than the addition method.

To Solve a System by Substitution:

1. Solve one equation for x or y, if necessary.
2. Substitute the expression from Step 1 into the other equation.
3. Solve the resulting equation.
4. Substitute the value from Step 3 into one of the original equations to find the other value.
5. Check, as necessary.

EXAMPLE 4 Solve each system.

a) $y = 2x - 1$

$6x - 2y = 4$

b) $x + 3y = -7$

$4x + 2y = 2$

SOLUTION 4

a) $y = 2x - 1$ — This equation is solved for y.

$6x - 2y = 4$	
$6x - 2(2x - 1) = 4$	Substitute $(2x - 1)$ for y.
$6x - 4x + 2 = 4$	Solve for x.
$2x + 2 = 4$	
$2x = 2$	Subtract 2 from both sides.
$x = 1$	Divide both sides by 2.

Now solve for y:

$y = 2x - 1$	
$y = 2(1) - 1$	Substitute $x = 1$.
$y = 2 - 1$	
$y = 1$	
$(1, 1)$	Proposed solution.

Try the check on your own. The solution to the system is (1, 1).

b) $x + 3y = -7$	This equation can be easily
$4x + 2y = 2$	solved for x.
$x + 3y = -7$	
$x + 3y - 3y = -7 - 3y$	Subtract $3y$ from both sides.
$x = -7 - 3y$	
$4x + 2y = 2$	Substitute into the other equation.
$4(-7 - 3y) + 2y = 2$	Substitute $(-7 - 3y)$ for x.
$-28 - 12y + 2y = 2$	Solve for y.
$-28 - 10y = 2$	Add similar terms.
$-28 - 10y + 28 = 2 + 28$	Add 28 to both sides.
$-10y = 30$	
$y = -3$	

Now solve for x:

$x + 3y = -7$	
$x + 3(-3) = -7$	Substitute $y = -3$.
$x - 9 = -7$	Add 9 to both sides.
$x = 2$	

Try the check on your own. The solution to the system is (2, –3).

8.2 SOLVING SYSTEMS OF LINEAR EQUATIONS IN THREE VARIABLES

Ordered triples

A linear equation in three variables is usually written with the variables x, y, and z, such as $2x - 3y + z = -12$. A solution to such an equation is called an ordered triple and written (x, y, z). For example, $(1, 4, -2)$ is a solution to $2x - 3y + z = -12$ since $2(1) - 3(4) + (-2) = -12$.

EXAMPLE 5 Determine whether the ordered triple is a solution to the equation $3x + 2y + 5z = 16$.

a) $(1, -1, 3)$

b) $(2, -1, -4)$

SOLUTION 5

a)

$$3x + 2y + 5z = 16$$

$$3(1) + 2(-1) + 5(3) \stackrel{?}{=} 16 \qquad \text{Substitute.}$$

$$3 - 2 + 15 \stackrel{?}{=} 16$$

$$16 = 16 \qquad \text{True.}$$

$(1, -1, 3)$ is a solution to $3x + 2y + 5z = 16$.

b)

$$3x + 2y + 5z = 16$$

$$3(2) + 2(-1) + 5(-4) \stackrel{?}{=} 16 \qquad \text{Substitute.}$$

$$6 - 2 - 20 \stackrel{?}{=} 16$$

$$-16 = 16 \qquad \text{False.}$$

$(2, -1, -4)$ is not a solution to $3x + 2y + 5z = 16$.

Solving systems

The graph of a linear equation in three variables is a plane (think of a desk top that has infinite length and width). Graphing in three dimensions is beyond our needs at this time. However, we can discuss the solution to a system of three linear equations in three unknowns, such as

$$\begin{aligned} x + y + z &= 4 \\ x - y + 2z &= -3 \\ 2x - y - 3z &= 4 \end{aligned}$$

If all three planes intersect in one point, the solution to the system is an ordered triple. We can use the addition technique from Section 1 to solve a system of three linear equations.

> **To Solve a System of Three Linear Equations:**
> 1. Label the equations *A*, *B*, *C*.
> 2. Choose a variable and eliminate it from two equations at a time.
> 3. Solve the resulting system of two equations in two unknowns.
> 4. Substitute the values from Step 3 into one of the original equations to find the third value.
> 5. Check your solution in all three equations.

EXAMPLE 6 Solve:

$$x + y + z = 4$$

$$x - y + 2z = -3$$

$$2x - y - 3z = 4$$

SOLUTION 6

$A\ \ x + y + z = 4$ Label the equations *A*, *B*, and *C*.

$B\ \ x - y + 2z = -3$

$C\ \ 2x - y - 3z = 4$

Let's eliminate *y*:

$A\ \ x + y + z = 4$ Add *A* and *B* to eliminate *y*.

$B\ \ \underline{x - y + 2z = -3}$

$2x + 3z = 1$

$A\ \ x + y + z = 4$ Add *A* and *C* to eliminate *y*.

$C\ \ \underline{2x - y - 3z = 4}$

$3x - 2z = 8$

Solve the resulting system:

$2x + 3z = 1$ Let's eliminate *z*.

$3x - 2z = 8$

$2(2x + 3z = 1) \Rightarrow 4x + 6z = 2$ Multiply by the coefficients of *z*.

$3(3x - 2z = 8) \Rightarrow 9x - 6z = 24$

$4x + 6z = 2$ Add the equations.

$\underline{9x - 6z = 24}$

$13x = 26$

$x = 2$ Solve for *x*.

$2x + 3z = 1$ Solve for *z*.

$2(2) + 3z = 1$ — Substitute $x = 2$.

$4 + 3z = 1$

$3z = -3$

$z = -1$

Now use $x = 2$ and $z = -1$ to find y:

$x + y + z = 4$ — Use one of the original equations.

$2 + y + (-1) = 4$

$y + 1 = 4$

$y = 3$

$(2, 3, -1)$ — Proposed solution.

Check:

Substitute $(2, 3, -1)$ into each of the original equations. $(2, 3, -1)$ is the solution to the system.

EXAMPLE 7 Solve.

$$6x + 4y - 2z = -7$$

$$3x - 2y + 8z = 7$$

$$x - 2y + 4z = 5$$

SOLUTION 7

$A \quad 6x + 4y - 2z = -7$

$B \quad 3x - 2y + 8z = 7$

$C \quad x - 2y + 4z = 5$

Label the equations A, B, and C.

Let's eliminate z:

$24x + 16y - 8z = -28 \quad (4 \cdot A)$ — Multiply 4 times equation A.

$\underline{3x - 2y + 8z = 7} \quad (B)$

$27x + 14y = -21$

$3x - 2y + 8z = 7 \quad (B)$ — Multiply -2 times equation C.

$\underline{-2x + 4y - 8z = -10} \quad (-2 \cdot C)$

$x + 2y = -3$

Solve the resulting system:

$$27x + 14y = -21$$
$$x + 2y = -3$$

Let's eliminate y.

$$27x + 14y = -21 \quad \Rightarrow \quad 27x + 14y = -21$$
$$-7(x + 2y = -3) \quad \Rightarrow \quad \underline{-7x - 14y = 21}$$
$$20x = 0$$

Solve for x.

$$x = 0$$

$$x + 2y = -3$$
$$0 + 2y = -3$$
$$y = -\frac{3}{2}$$

Substitute $x = 0$.

Now we use $x = 0$ and $y = -\frac{3}{2}$ to find z:

$$6x + 4y - 2z = -7$$

Use one of the original equations.

$$6(0) + 4\left(-\frac{3}{2}\right) - 2z = -7$$

Substitute.

$$0 - 6 - 2z = -7$$
$$-2z = -1$$

Add 6 to both sides.

$$z = \frac{1}{2}$$

Divide by -2.

Check:

Substitute $(0, -\frac{3}{2}, \frac{1}{2})$ into each of the original equations. $(0, -\frac{3}{2}, \frac{1}{2})$ is the solution to the system.

Missing terms

If one or more equations are missing a term, we can alter step 2 in the process.

EXAMPLE 8 Solve.

$$2x - y + 3z = 0$$

$$-3y + 2z = -5$$

$$-5x + 3z = -13$$

SOLUTION 8

A $\quad 2x - y + 3z = 0$

B $\quad -3y + 2z = -5$

C $\quad -5x + 3z = -13$

If we choose to eliminate x, we only need to use equations A and C since equation B is already missing an x term.

$10x - 5y + 15z = \quad 0 \quad (5 \cdot A)$	Multiply 5 times equation A.
$\underline{-10x \qquad + 6z = -26} \quad (2 \cdot C)$	Multiply 2 times equation C.
$-5y + 21z = -26$	Add the equations together.

Now solve the system:

$-3y + \ 2z = -5$	Equation B.
$-5y + 21z = -26$	New equation with x term eliminated.

$$5(-3y + \ 2z = -5) \Rightarrow -15y + 10z = -25 \qquad \text{Solve for } z.$$

$$-3(-5y + 21z = -26) \Rightarrow \underline{15y - 63z = 78}$$

$$53z = 53$$

$$z = -1$$

$-3y + 2z = -5$	Solve for y.
$-3y + 2(-1) = -5$	Substitute $z = -1$.
$-3y - 2 = -5$	
$-3y = -3$	
$y = 1$	

Now use $y = 1$ and $z = -1$ to find x:

$2x - y + 3z = 0$	
$2x - (1) + 3(-1) = 0$	Substitute $y = 1$ and $z = -1$.
$2x - 4 = 0$	Add similar terms.
$2x = 4$	Add 4 to both sides.
$x = 2$	Divide both sides by 2.

Check:

Substitute $(2, 1, -1)$ into each of the original equations. $(2, 1, -1)$ is the solution to the system.

Inconsistent systems and dependent equations

Recall that the solution to a system of two linear equations in two unknowns consists of one point, an infinite number of points (dependent equations), or no point (inconsistent system). A similar situation exists with three linear equations in three unknowns. In the process of solving a system, all the variables may be eliminated. If the resulting statement is true, the equations are dependent, which means at least two of the planes coincide. In this case we say that the solution is not unique (it is either a line or a plane). If the resulting statement is false, the system is inconsistent and the solution set is empty.

EXAMPLE 9 Solve.

a) $3x + 2y + z = 4$

$9x + 6y + 3z = 12$

$6x + 4y + 2z = 8$

b) $5x - 3y + 2z = 4$

$15x - 9y + 6z = 8$

$2x - 6y + 3z = 12$

SOLUTION 9

a) $A \quad 3x + 2y + z = 4$
$B \quad 9x + 6y + 3z = 12$
$C \quad 6x + 4y + 2z = 8$

Label the equations A, B, and C.

Let's eliminate z:

$-9x - 6y - 3z = -12 \quad (-3 \cdot A)$

Multiply -3 times equation A.

$\underline{9x + 6y + 3z = 12} \quad (B)$

$0 = 0$

Add the equations.

Since all variables were eliminated, and the resulting statement is true, equations A and B are dependent and there is no unique solution to the system. If you work with equations A and C, you will find that they are also dependent equations which means all three planes coincide.

b) $A \quad 5x - 3y + 2z = 4$
$B \quad 15x - 9y + 6z = 8$
$C \quad 2x - 6y + 3z = 12$

Label the equations A, B, and C.

Let's eliminate x:

$-15x + 9y - 6z = -12$ $(-3 \cdot A)$ — Multiply -3 times equation A.

$\underline{15x - 9y + 6z = \quad 8}$ (B)

$0 = -4$ — Add the equations.

Since all variables were eliminated and the resulting statement is false, the system is inconsistent and the solution set is empty, $\varnothing$.

8.3 APPLICATIONS FOR SOLVING EQUATIONS

In this section we will solve several types of word problems that involve two unknowns and three unknowns. In some cases it may help to organize the data in a table; in other cases a carefully labeled picture may help. The following list will also help you solve word problems.

1. State what each variable you are using represents.
2. Translate the problem into a system of equations.
3. Solve the system by addition or subtraction.
4. Provide answers for the variable(s) as required.
5. Check your answers(s) in the original word problem.

Number applications

First let's work a problem that can be solved with two equations.

EXAMPLE 10 The sum of two times the larger of two numbers and six times the smaller is 34. The difference between three times the larger and four times the smaller is 12. Find both numbers.

SOLUTION 10

Let

$x =$ the larger number — State what each variable

$y =$ the smaller number — represents.

$2x + 6y = 34$ — Write a system of equations.

$3x - 4y = 12$

Let's eliminate y:

$2(2x + 6y = 34) \Rightarrow 4x + 12y = 68$

$3(3x - 4y = 12) \Rightarrow \underline{9x - 12y = 36}$

$13x = 104$ Add the equations.

$x = 8$ Divide by 13.

$2x + 6y = 34$ Use an original equation to find y.

$2(8) + 6y = 34$ Substitute $x = 8$.

$16 + 6y = 34$ Multiply.

$6y = 18$ Subtract 16 from both sides.

$y = 3$ Divide by 6.

The larger number is 8 and the smaller number is 3.

Check:

Two times the larger and six times the smaller is 34. The difference between three times the larger and four times the smaller is 12.

We can also solve number problems that require three equations.

EXAMPLE 11 The sum of three numbers is 6. The sum of two times the first number and three times the second number is 7 more than the third number. Three times the first number is 6 more than the sum of the second and third number. Find all three numbers.

SOLUTION 11

Let

$x =$ the first number

$y =$ the second number

$z =$ the third number

State what each variable represents.

$x + y + z = 6$ Write a system of equations.

$2x + 3y = 7 + z$

$3x = 6 + y + z$

To solve the system, first write each equation in standard form ($ax + by + cz = d$).

A $x + y + z = 6$

B $2x + 3y - z = 7$

C $3x - y - z = 6$

Label the equations A, B, and C.

Let's eliminate z:

$$x + y + z = 6 \quad (A)$$
$$\underline{2x + 3y - z = 7} \quad (B)$$
$$3x + 4y = 13$$

Add the equations.

$$x + y + z = 6 \quad (A)$$
$$\underline{3x - y - z = 6} \quad (C)$$
$$4x = 12$$

Both y and z were eliminated.

$x = 3$	Solve for x.
$3x + 4y = 13$	Solve for y.
$3(3) + 4y = 13$	Substitute $x = 3$.
$9 + 4y = 13$	Multiply.
$4y = 4$	Subtract 9 from both sides.
$y = 1$	Divide by 4.
$x + y + z = 6 \; (A)$	Use one of the original equations to find z.

$$3 + 1 + z = 6$$
$$4 + z = 6$$
$$z = 2$$

The numbers are 3, 1, and 2. Check the solution in each of the original equations.

Money applications

Systems of equations can also be used to solve money applications. To find the total value in this type of problem, multiply the quantity times the value of each item. For example, if you purchase 10 tickets for \$7.50 each, the total value of the tickets sold is \$7.50(10) = \$75.00.

EXAMPLE 12 Tickets for the Labor Day picnic cost \$3.50 for adults and \$2.00 for children. A total of 600 tickets were sold for \$1800. How many of each type of ticket were sold?

SOLUTION 12

Let

x = the number of adult tickets sold

y = the number of childrens tickets sold

State what each variable represents.

A table may help organize the data:

	Adult tickets	Children's tickets	Total
Number	x	y	600
Value	$3.50x$	$2.00y$	1800

We can write two equations:

$x + y = 600$ — 600 tickets were sold.

$3.50x + 2.00y = 1800$ — $1800 represents the total value.

Let's solve by substitution.

$x = 600 - y$ — Solve the first equation for x.

$3.50(600 - y) + 2.00y = 1800$ — Substitute into the second equation.

$2100 - 3.50y + 2.00y = 1800$ — Multiply.

$2100 - 1.50y = 1800$ — Combine similar terms.

$-1.50y = -300$ — Subtract 2100 from both sides.

$y = 200$ — Divide by -1.50.

$x = 600 - y$ — Solve for x.

$x = 600 - 200$ — Substitute $y = 200$.

$x = 400$

There were 400 adult tickets sold and 200 children tickets. Check by substituting the solution in the original equations.

EXAMPLE 13 Bill's coin collection consists of 155 coins with a total value of $15.50. If the coins are nickels, dimes, and quarters, and twice the number of dimes is 70 more than the sum of the number of nickels and quarters. How many of each coin are in the collection?

SOLUTION 13

Let

x = the number of nickels — State what each variable represents.

y = the number of dimes

z = the number of quarters

A table may help organize the data:

	Nickels	Dimes	Quarters	Total
Number	x	y	z	155
Value	$0.05x$	$0.10y$	$0.25z$	15.50

The table provides data for two equations:

$x + y + z = 155$ — The total number of coins is 155.

$0.05x + 0.10y + 0.25z = \$15.50$ — The total value is \$15.50.

$2y = 70 + x + z$ — The third equation is the translation of the phrase "twice" the number of dimes is 70 more than the sum of the number of nickels and quarters."

Now write the equations in the form $ax + by + cz = d$:

$$
\begin{array}{llcl}
A & x + y + z & = & 155 \\
B & 5x + 10y + 25z & = & 1550 \\
C & -x + 2y - z & = & 70
\end{array}
$$

Let's eliminate x:

$-5x - 5y - 5z = 155 \quad (-5 \bullet A)$ — Multiply equation A by -5.

$\underline{5x + 10y + 25z = 1550} \quad (B)$

$5y + 20z = 775$ — Add A and B.

$x + y + z = 155 \quad (A)$

$\underline{-x + 2y - z = 70} \quad (C)$

$3y = 225$ — Add A and C.

$y = 75$ — Solve for y.

$5y + 20z = 775$ — Solve for z.

$5(75) + 20z = 775$

$375 + 20z = 775$

$20z = 400$

$z = 20$

$x + y + z = 155$ — Use one of the original equations to solve for x.

$x + 75 + 20 = 155$ — Substitute $y = 75$ and $z = 20$.

$x + 95 = 155$ — Subtract 95 from both sides.

$x = 60$

There are 60 nickels ($3.00), 75 dimes ($7.50), and 20 quarters ($5.00). Thus there are 155 coins with a total value of $15.50 and twice the number of dimes is 70 more than the sum of the number of nickels and quarters.

Interest

Interest problems are money problems with the value stated as a percent. Remember to change percents to decimals.

EXAMPLE 14 Sarah invested a total of $8000 in two accounts. One account earns 6% interest annually and the other earns 8% interest annually. If the total interest earned from both accounts in one year is $600, find the amount invested in each account.

SOLUTION 14

Let

x = the amount of money invested at 6%

y = the amount of money invested at 8%

$x + y = 8000$ — Total investment was $8000.

$0.06x + 0.08y = 600$ — Interest from both accounts was $600.

Now solve the system.

$-0.06x - 0.06y = -480$ — Multiply by –.06.

$0.06x + 0.08y = 600$

$0.02y = 120$ — Add the equations.

$y = 6000$ — Divide by 0.02.

$x + y = 8000$ — Use one of the original equations to find x.

$$x + 6000 = 8000$$
$$x = 2000$$

The amount of money invested at 6% was \$2000 and the amount of money invested at 8% was \$6000.

$D = RT$ applications

Systems of equations can also be used to solve applications that involve the formula

$$D = RT$$
Distance = Rate times Time

EXAMPLE 15 It takes 3 hours for a boat to travel 48 miles downstream. The same boat can travel 40 miles upstream in 5 hours. What is the rate of the boat in still water and the speed of the current?

SOLUTION 15

Let

x = the speed of the boat in still water

y = the speed of the current

A table may help organize the data:

	R	T	D
Downstream	$x + y$	3	48
Upstream	$x - y$	5	40

Note that we use $x + y$ for the rate downstream since the boat is traveling *with* the current. Traveling against the current upstream we subtract the speed of the current from the speed of the boat in still water.

We can write two equations from this data:

$3(x + y) = 48$ Use $RT = D$.

$5(x - y) = 40$

Solve the system

$3x + 3y = 48$

$5x - 5y = 40$

by eliminating y:

$$\begin{aligned} 5(3x + 3y = 48) \quad &=> \quad 15x + 15y = 240 \\ 3(5x - 5y = 40) \quad &=> \quad \underline{15x - 15y = 120} \\ & \quad\quad 30x \quad\quad = 360 \quad \text{Add the equations.} \\ & \quad\quad\quad\quad\quad x = 12 \end{aligned}$$

$$\begin{aligned} 3x + 3y &= 48 && \text{Now solve for } y. \\ 3(12) + 3y &= 48 && \text{Substitute } x = 12. \\ 36 + 3y &= 48 \\ 3y &= 12 \\ y &= 4 \end{aligned}$$

The boat travels 12 miles per hour in still water, and the speed of the current is 4 miles per hour.

8.4 DETERMINANTS

In Sections 8.1 and 8.2 we solved systems of equations by addition and substitution. Another method for solving systems uses determinants. In this section we will find the value of determinants, and then use these values to solve systems in section 8.5.

Definition. The value of the determinant $\begin{vmatrix} a & b \\ c & d \end{vmatrix}$ is $ad - bc$.

A **determinant** is an array of numbers written between vertical bars. The array of numbers must have an equal number of rows and columns (called a square array). In our definition, we've used a 2 x 2 (read "2 by 2") array (2 rows and 2 columns).

EXAMPLE 16 Find the value of each determinant.

a) $\begin{vmatrix} 2 & 5 \\ 3 & 9 \end{vmatrix}$

b) $\begin{vmatrix} 0 & -1 \\ 2 & 6 \end{vmatrix}$

SOLUTION 16

a) $\begin{vmatrix} 2 & 5 \\ 3 & 9 \end{vmatrix}$

$= 2(9) - 5(3)$ Use $ad - bc$.

$= 18 - 15$ Multiply.

$= 3$

b) $\begin{vmatrix} 0 & -1 \\ 2 & 6 \end{vmatrix}$

$= 0(6) - (-1)2$ Use $ad - bc$.

$= 0 - (-2)$ Multiply.

$= 2$

3 x 3 determinants

The value of 3 x 3 determinants, read "3 by 3" (3 rows and 3 columns) can be found by either of two methods. Note that the first method can *only* be used to find the value of a 3 x 3 determinant, while the second method can be used to find the value of a 3 x 3 determinant as well as a 4 x 4 or a larger determinant.

Method 1

The following steps are used to find the value of a 3 x 3 determinant.

1. Write the determinant with the first two columns repeated on the right.
2. Multiply the elements down the three full diagonals and add them together.
3. Multiply the elements up the three full diagonals and subtract them.
4. Add the answers from Step 2 and 3 to find the value of the determinant.

EXAMPLE 17 Find the value of the determinant.

$$\begin{vmatrix} 1 & 2 & 2 \\ 2 & 3 & 3 \\ 1 & -1 & -2 \end{vmatrix}$$

SOLUTION 17

Repeat the first two columns on the right.

$$\begin{matrix} 1 & 2 & 2 & 1 & 2 \\ 2 & 3 & 3 & 2 & 3 \\ 1 & -1 & -2 & 1 & -1 \end{matrix}$$

Multiply down the three full diagonals.

$$\begin{matrix} 1 & 2 & 2 & 1 & 2 \\ 2 & 3 & 3 & 2 & 3 \\ 1 & -1 & -2 & 1 & -1 \\ & & + & + & + \end{matrix}$$

Add the products.

$$+1(3)(-2) + 2(3)(1) + 2(2)(-1) =$$
$$-6 \quad + \quad 6 \quad - \quad 4 \quad = -4$$

Multiply up the three full diagonals.

$$\begin{matrix} & & - & - & - \\ 1 & 2 & 2 & 1 & 2 \\ 2 & 3 & 3 & 2 & 3 \\ 1 & -1 & -2 & 1 & -1 \end{matrix}$$

Subtract the products.

$$-1(3)(2) - (-1)(3)(1) - (-2)(2)(2) =$$
$$-6 \quad + \quad 3 \quad + \quad 8 \quad = 5$$

Add the answers from steps 2 and 3.

$$-4 + 5 = 1$$

The value of the determinant is 1.

EXAMPLE 18 Find the value of the determinant.

$$\begin{vmatrix} 0 & 2 & 2 \\ 1 & 3 & 3 \\ 2 & -1 & -2 \end{vmatrix}$$

SOLUTION 18

Repeat the first two columns on the right.

$$\begin{matrix} 0 & 2 & 2 & 0 & 2 \\ 1 & 3 & 3 & 1 & 3 \\ 2 & -1 & -2 & 2 & -1 \end{matrix}$$

$$\begin{matrix} 0 & 2 & 2 & 0 & 2 \\ 1 & 3 & 3 & 1 & 3 \\ 2 & -1 & -2 & 2 & -1 \end{matrix}$$

Multiply down the three full diagonals.

$+ \quad + \quad +$

$0(3)(-2) + 2(3)(2) + (2)(1)(-1) =$

$0 \quad + \quad 12 \quad + \quad -2 \quad = 10$

Add the products.

$- \quad - \quad -$

$$\begin{matrix} 0 & 2 & 2 & 0 & 2 \\ 1 & 3 & 3 & 1 & 3 \\ 2 & -1 & -2 & 2 & -1 \end{matrix}$$

Multiply up the three full diagonals.

$-(2)(3)(2) - (-1)(3)(0) - (-2)(1)(2) =$

$-12 \quad - \quad 0 \quad + \quad 4 \quad = -8$

Subtract the products.

$10 - 8 = 2$

Add the answers from steps 2 and 3.

The value of the determinant is 2.

Method 2
Expansion by minors

Method 2 for finding the value of a determinant is called expansion by minors.

> **Definition.** The **minor** of an element is the determinant formed when the row and column containing the element are deleted.

EXAMPLE 19 Find the minor of the element in the first row and first column of

$$\begin{vmatrix} 1 & 2 & 2 \\ 2 & 3 & 3 \\ 1 & -1 & -2 \end{vmatrix}$$

SOLUTION 19

$$\begin{vmatrix} \textcircled{1} & 2 & 2 \\ 2 & 3 & 3 \\ 1 & -1 & -2 \end{vmatrix}$$

1 is the element in the first row and first column.

$$\begin{vmatrix} 1 & 2 & 2 \\ 2 & 3 & 3 \\ 1 & -1 & -2 \end{vmatrix}$$

Delete the row and column that contains 1.

$$\begin{vmatrix} 3 & 3 \\ -1 & -2 \end{vmatrix}$$

The remaining determinant is the minor of the element in the first row and first column.

EXAMPLE 20 Find the minor of the element in the first row and second column of

$$\begin{vmatrix} 1 & 2 & 2 \\ 2 & 3 & 3 \\ 1 & -1 & -2 \end{vmatrix}$$

SOLUTION 20

$$\begin{vmatrix} 1 & \textcircled{2} & 2 \\ 2 & 3 & 3 \\ 1 & -1 & -2 \end{vmatrix}$$

2 is the element in the first row and second column.

$$\begin{vmatrix} 1 & 2 & 2 \\ 2 & 3 & 3 \\ 1 & -1 & -2 \end{vmatrix}$$

Delete the row and column containing 2.

$$\begin{vmatrix} 2 & 3 \\ 1 & -2 \end{vmatrix}$$

The remaining determinant is the minor of the element in the first row and second column.

Signs for expansion

We will multiply an element times its minor times +1 or –1 to find the value of a determinant. The following table indicates whether to multiply by +1 or –1:

$$\begin{vmatrix} + & - & + \\ - & + & - \\ + & - & + \end{vmatrix}$$

Expansion across a row or column

The value of the determinant can be found using the following steps:

1. Choose a row or column to expand across.
2. Multiply each element in that row or column times its minor times +1 or −1 (see the previous sign table.)
3. Evaluate each product and find the sum.

EXAMPLE 21 Find the value of the determinant by expanding across row 1.

$$\begin{vmatrix} 1 & 2 & 2 \\ 2 & 3 & 3 \\ 1 & -1 & -2 \end{vmatrix}$$

SOLUTION 21

In Examples 19 and 20 we found the minors for the first two elements in row
1.

Let's find the minor for the element in row 1 and column 3.

$$\begin{vmatrix} 1 & 2 & \textcircled{2} \\ 2 & 3 & 3 \\ 1 & -1 & -2 \end{vmatrix}$$

Delete the row and column containing 2.

$$\begin{vmatrix} 2 & 3 \\ 1 & -1 \end{vmatrix}$$

The remaining determinant is the minor of the element in the first row and third column.

Find sign in sign table

$$(+1)(1)\begin{vmatrix} 3 & 3 \\ -1 & -2 \end{vmatrix} + (-1)(2)\begin{vmatrix} 2 & 3 \\ 1 & -2 \end{vmatrix} + (+1)(2)\begin{vmatrix} 2 & 3 \\ 1 & -1 \end{vmatrix} =$$

$1(-6-(-3)) + -2(-4-3) + 2(-2-3)$	Evaluate each determinant.
$1(-3) + -2(-7) + 2(-5) =$	Simplify inside parentheses.
$-3 + 14 - 10 =$	Add.
1	Value of the determinant.

Return to Example 17 to compare this method and answer to method 1.

EXAMPLE 22 Find the value of the determinant

$$\begin{vmatrix} 0 & 2 & 2 \\ 1 & 3 & 3 \\ 2 & -1 & -2 \end{vmatrix}$$

by expanding down column 2.

SOLUTION 22

Use the elements in column 2 and the signs for column 2 in the sign table to write:

$$\begin{vmatrix} 0 & 2 & 2 \\ 1 & 3 & 3 \\ 2 & -1 & -2 \end{vmatrix}$$

$$(-1)(2)\begin{vmatrix} 1 & 3 \\ 2 & -2 \end{vmatrix} + (+1)(3)\begin{vmatrix} 0 & 2 \\ 2 & -2 \end{vmatrix} + (-1)(-1)\begin{vmatrix} 0 & 2 \\ 1 & 3 \end{vmatrix} =$$

$-2(-2-6) + 3(0-4) + 1(0-2) =$	Evaluate each determinant.
$-2(-8) + 3(-4) + 1(-2) =$	Simplify inside parentheses.
$16 - 12 - 2 =$	Add.
2	Value of the determinant.

Compare this result to Example 18.

The determinant can be found by expanding along any row or column. You may want to rework Example 21 or 22 by expanding along a different row or column.

8.5 SOLVING SYSTEMS OF LINEAR EQUATIONS USING CRAMER'S RULE

Cramer's rule for a 2 x 2

Cramer's Rule uses determinants to solve for x and y in a system of two equations in two unknowns (2 x 2).

Cramer's Rule. The solution to the system

$$a_1x + b_1y = c$$
$$a_2x + b_2y = d$$

is $x = \dfrac{\begin{vmatrix} c & b_1 \\ d & b_2 \end{vmatrix}}{\begin{vmatrix} a_1 & b_1 \\ a_2 & b_2 \end{vmatrix}}$ and $y = \dfrac{\begin{vmatrix} a_1 & c \\ a_2 & d \end{vmatrix}}{\begin{vmatrix} a_1 & b_1 \\ a_2 & b_2 \end{vmatrix}}$

EXAMPLE 23 Solve using Cramer's Rule.

$$x + 5y = 3$$
$$3x - 2y = 9$$

SOLUTION 23

Set up the determinants:

$$x = \frac{\begin{vmatrix} 3 & 5 \\ 9 & -2 \end{vmatrix}}{\begin{vmatrix} 1 & 5 \\ 3 & -2 \end{vmatrix}} \text{ and } y = \frac{\begin{vmatrix} 1 & 3 \\ 3 & 9 \end{vmatrix}}{\begin{vmatrix} 1 & 5 \\ 3 & -2 \end{vmatrix}}$$

Now evaluate each determinant:

$$x = \frac{3(-2) - 9(5)}{1(-2) - 3(5)} \qquad y = \frac{1(9) - 3(3)}{1(-2) - 3(5)}$$

$$x = \frac{-6 - 45}{-2 - 15} \qquad y = \frac{9 - 9}{-2 - 15}$$

$$x = \frac{-51}{-17} \qquad y = \frac{0}{-17}$$

$$x = 3 \qquad y = 0$$

Note that the determinant in the denominator is the same for x and y, and so can be computed once.

EXAMPLE 24 Solve using Cramer's Rule.

$2x - 3y = 4$

$3x - 4y = 5$

SOLUTION 24

Set up the determinants:

$$x = \frac{\begin{vmatrix} 4 & -3 \\ 5 & -4 \end{vmatrix}}{\begin{vmatrix} 2 & -3 \\ 3 & -4 \end{vmatrix}} \text{ and } y = \frac{\begin{vmatrix} 2 & 4 \\ 3 & 5 \end{vmatrix}}{\begin{vmatrix} 2 & -3 \\ 3 & -4 \end{vmatrix}}$$

Now evaluate each determinant:

$$x = \frac{4(-4) - 5(-3)}{2(-4) - 3(-3)} \qquad y = \frac{2(5) - 3(4)}{2(-4) - 3(-3)}$$

$$x = \frac{-16 + 15}{-8 + 9} \qquad y = \frac{10 - 12}{1}$$

$$x = \frac{-1}{1} \qquad y = \frac{-2}{1}$$

$$x = -1 \qquad y = -2$$

Cramer's rule for a 3 x 3

We can also use Cramer's rule to solve for x, y, and z in a system of three equations and three unknowns:

Cramer's Rule: The solution to the system

$$a_1x + b_1y + c_1z = d$$
$$a_2x + b_2y + c_2z = e$$
$$a_3x + b_3y + c_3z = f$$

$$x = \frac{\begin{vmatrix} d & b_1 & c_1 \\ e & b_2 & c_2 \\ f & b_3 & c_3 \end{vmatrix}}{\begin{vmatrix} a_1 & b_1 & c_1 \\ a_2 & b_2 & c_2 \\ a_3 & b_3 & c_3 \end{vmatrix}} \qquad y = \frac{\begin{vmatrix} a_1 & d & c_1 \\ a_2 & e & c_2 \\ a_3 & f & c_3 \end{vmatrix}}{\begin{vmatrix} a_1 & b_1 & c_1 \\ a_2 & b_2 & c_2 \\ a_3 & b_3 & c_3 \end{vmatrix}} \qquad z = \frac{\begin{vmatrix} a_1 & b_1 & d \\ a_2 & b_2 & e \\ a_3 & b_3 & f \end{vmatrix}}{\begin{vmatrix} a_1 & b_1 & c_1 \\ a_2 & b_2 & c_2 \\ a_3 & b_3 & c_3 \end{vmatrix}}$$

EXAMPLE 25 Solve using Cramer's Rule.

$$x + 2y + 2z = 0$$
$$2x + 3y + 3z = 1$$
$$x - y - 2z = 2$$

SOLUTION 25

Set up the determinants for x:

$$x = \frac{\begin{vmatrix} 0 & 2 & 2 \\ 1 & 3 & 3 \\ 2 & -1 & -2 \end{vmatrix}}{\begin{vmatrix} 1 & 2 & 2 \\ 2 & 3 & 3 \\ 1 & -1 & -2 \end{vmatrix}}$$

We previously found each of these determinants (the numerator in Example 22 and the denominator in Example 21), so

$$x = \frac{2}{1} = 2$$

Set up the determinant for the numerator of y, and use the value we previously found for the denominator:

$$y = \frac{\begin{vmatrix} 1 & 0 & 2 \\ 2 & 1 & 3 \\ 1 & 2 & -2 \end{vmatrix}}{1}$$

Expand along row 1 of the determinant in the numerator.

$$+1(1)\begin{vmatrix} 1 & 3 \\ 2 & -2 \end{vmatrix} + (-1)(0)\begin{vmatrix} 2 & 3 \\ 1 & -2 \end{vmatrix} + (+1)(2)\begin{vmatrix} 2 & 1 \\ 1 & 2 \end{vmatrix} =$$

$= 1(-2 - 6) + 0(-4 - 3) + 2(4 - 1)$ Evaluate each determinant.

$= 1(-8) + 0 + 2(3)$ Simplify inside parentheses.

$= -8 + 6$ Multiply.

$= -2$

So $y = \frac{-2}{1} = -2$

Find z:

$$z = \frac{\begin{vmatrix} 1 & 2 & 0 \\ 2 & 3 & 1 \\ 1 & -1 & 2 \end{vmatrix}}{1}$$

Expand along row 1.

$$+1(1)\begin{vmatrix} 3 & 1 \\ -1 & 2 \end{vmatrix} + (-1)(2)\begin{vmatrix} 2 & 1 \\ 1 & 2 \end{vmatrix} + (+1)(0)\begin{vmatrix} 2 & 3 \\ 1 & -1 \end{vmatrix} =$$

$= 1(6 - (-1)) + (-2)(4 - 1) + 0(-2 - 3)$ Evaluate each determinant.

$= 1(7) + (-2)(3) + 0$ Simplify inside parentheses.

$= 7 - 6$ Multiply.

$= 1$

So $z = \frac{1}{1} = 1$

The solution to the system is (2, –2, 1).

Note that each determinant was found using Method 2, and expanding along row 1. Either method could have been used to find the determinants. Also, if the determinant in the denominator should equal 0, Cramer's Rule does *not* apply, and an alternate method should be used to solve the system.

8.6 SOLVING SYSTEMS OF LINEAR EQUATIONS USING MATRICES

So far in this chapter we have solved systems of equations using addition, substitution, and Cramer's Rule. In this section we will present another method for solving systems using matrices.

Matrices

A **matrix** is an array of numbers, such as

$$\begin{vmatrix} 1 & 5 \\ 3 & -2 \end{vmatrix}$$

We can represent a system of equations using an augmented matrix, where a bar separates the
coefficients of the variables from the constant terms.

For example, the system

$$\begin{aligned} 3x + 2y &= 8 \\ x - y &= -3 \end{aligned}$$

is represented by the augmented matrix

$$\left[\begin{array}{cc|c} 3 & 2 & 8 \\ 1 & -1 & -3 \end{array}\right]$$

We write the coefficients of x and y to the left of the bar, and the constants to the right of the bar.

EXAMPLE 26 Write the augmented matrix for each system of equations.

a) $x + 5y = 3$

$3x - 2y = 9$

b) $x + 2y + 2z = 0$

$2x + 3y + 3z = 1$

$x - y - 2z = 2$

SOLUTION 26

a) $\left[\begin{array}{cc|c} 1 & 5 & 3 \\ 3 & -2 & 9 \end{array}\right]$

Write the coefficients on the left and the constants on the right.

b) $\left[\begin{array}{ccc|c} 1 & 2 & 2 & 0 \\ 2 & 3 & 3 & 1 \\ 1 & -1 & -2 & 2 \end{array}\right]$

EXAMPLE 27 Convert each augmented matrix into a system of equations.

a) $\left[\begin{array}{cc|c} 1 & 0 & 3 \\ 0 & 1 & 0 \end{array}\right]$

b) $\left[\begin{array}{ccc|c} 1 & 0 & 0 & 2 \\ 0 & 1 & 0 & -2 \\ 0 & 0 & 1 & 1 \end{array}\right]$

SOLUTION 27

a) $x + 0 \cdot y = 3$ or $x = 3$

$0 \cdot x + y = 0$ or $y = 0$

b) $x + 0 \cdot y + 0 \cdot z = 2$ or $y = 2$

$0 \cdot x + y + 0 \cdot z = -2$ or $y = -2$

$0 \cdot x + 0 \cdot y + z = 1$ or $z = 1$

As the previous example demonstrates, the solution to a system written in augmented matrix form can be read from the constants on the right if the coefficients on the left consist of 1's and 0's in the patterns

$$\left[\begin{array}{cc|c} 1 & 0 & a \\ 0 & 1 & b \end{array}\right] \qquad \left[\begin{array}{ccc|c} 1 & 0 & 0 & c \\ 0 & 1 & 0 & d \\ 0 & 0 & 1 & e \end{array}\right]$$

Our goal will be to perform matrix row operations on the augmented matrix to convert the left side into 1's and 0's in these patterns.

Matrix Row Operations
1. Any two rows may be interchanged.
2. Any row may be multiplied by a nonzero constant.
3. Any row may be added to a nonzero multiple of another row.

EXAMPLE 28 Solve using matrices.

$x + 5y = 3$

$3x - 2y = 9$

SOLUTION 28

$$\left[\begin{array}{cc|c} 1 & 5 & 3 \\ 3 & -2 & 9 \end{array}\right]$$

Write the augmented matrix.

We want the left side to become $\begin{bmatrix} 1 & 0 \\ 0 & 1 \end{bmatrix}$

It's usually easier to produce 0's. Let's multiply row 1 by –3 and add to row
2:

$$\begin{array}{rrrr} -3{\cdot}R_1 & -3 & -15 & -9 \\ R_2 & 3 & -2 & 9 \\ & 0 & -17 & 0 \end{array}$$

Now replace row 2 with 0 –17 0:

$$\left[\begin{array}{cc|c} 1 & 5 & 3 \\ 0 & -17 & 0 \end{array}\right]$$

$$\begin{array}{rrrr} 17{\cdot}R_1 & 17 & 85 & 1 \\ 5{\cdot}R_2 & 0 & -85 & 0 \\ & 17 & 0 & 51 \end{array}$$

Multiply 17 times row 1.
Multiply 5 times row 2.
Add the rows.

Now replace row 1 with 17 0 51:

$$\left[\begin{array}{cc|c} 7 & 0 & 51 \\ 0 & -17 & 0 \end{array}\right]$$

We can now multiply row 1 by $\frac{1}{17}$ and row 2 by $-\frac{1}{17}$:

$$\begin{array}{rrrr} \frac{1}{17}{\cdot}R_1 & 1 & 0 & 3 \\ -\frac{1}{17}{\cdot}R_2 & 0 & 1 & 0 \end{array}$$

and the augmented matrix becomes

$$\left[\begin{array}{cc|c} 1 & 0 & 3 \\ 0 & 1 & 0 \end{array}\right]$$

The solution to the system is
$x = 3, y = 0$ or $(3, 0)$.

EXAMPLE 29 Solve using matrices.

$$x + 2y + 2z = 0$$

$$2x + 3y + 3z = 1$$

$$x - y - 2z = 2$$

SOLUTION 29

Write the augmented matrix.

$$\left[\begin{array}{ccc|c} 1 & 2 & 2 & 0 \\ 2 & 3 & 3 & 1 \\ 1 & -1 & -2 & 2 \end{array}\right]$$

We want the left side to become $\begin{bmatrix} 1 & 0 & 0 \\ 0 & 1 & 0 \\ 0 & 0 & 1 \end{bmatrix}$

Again, it's usually easier to produce 0's. We'll show the work above, and the new augmented matrix below.

R_2	2	3	3	1
$-2 \cdot R_3$	–2	2	4	–4
	0	5	7	–3

$$\left[\begin{array}{ccc|c} 1 & 2 & 2 & 0 \\ 2 & 3 & 3 & 1 \\ 0 & 5 & 7 & -3 \end{array}\right]$$

$-2 \cdot R_1$	–2	–4	–4	0
R_2	2	3	3	1
	0	–1	–1	1

$$\left[\begin{array}{ccc|c} 1 & 2 & 2 & 0 \\ 0 & -1 & -1 & 1 \\ 0 & 5 & 7 & -3 \end{array}\right]$$

$5 \cdot R_1$	0	–5	–5	5
R_3	0	5	7	–3
	0	0	2	2

$$\left[\begin{array}{ccc|c} 1 & 2 & 2 & 0 \\ 0 & -1 & -1 & 1 \\ 0 & 0 & 2 & 2 \end{array}\right]$$

We can actually stop the process here and solve the system of equations

$$\begin{aligned} x + 2y + 2z &= 0 \\ -y - z &= 1 \\ 2z &= 2 \end{aligned}$$

Beginning at the bottom (this method is called backward substitution),

$2z = 2$

$z = 1$ Divide by 2.

Then use the second equation to find y:

$-y - z = 1$

$-y - (1) = 1$ Substitute $z = 1$.

$-y = 2$ Add 1 to both sides.

$y = -2$ Divide by -1.

Finally, use the top equation to find x:

$x + 2y + 2z = 0$

$x + 2(-2) + 2(1) = 0$ Substitute $y = -2$ and $z = 1$.

$x - 4 + 2 = 0$ Multiply.

$x - 2 = 0$ Add similar terms.

$x = 2$ Add 2 to both sides.

The solution is (2, –2, 1). Compare this technique and solution to Example 25 where we solved this system using Cramer's Rule.Note that we could have continued to use matrix row operations until the left side fit the pattern

$$\begin{matrix} 1 & 0 & 0 \\ 0 & 1 & 0 \\ 0 & 0 & 1 \end{matrix}$$

However, it is generally faster to produce 0's in the lower corner, and complete the problem using backward substitution.

Practice Exercises

1. Solve each system by graphing.

(a) $x + y = 3$

$2x - y = 0$

(b) $2x - 3y = -9$

$x + y = -2$

2. Solve each system

(a) $2x + y = 7$

$x - y = 5$

(b) $3x + 2y = 4$

$4x - 5y = -33$

(c) $\frac{1}{2}x - \frac{2}{3}y = 1$

$\frac{1}{4}x + \frac{1}{2}y = -2$

(d) $2x + y = 4$

$4x + 3y = 11$

3. Determine whether the following systems are dependent or inconsistent.

(a) $6x + 5y = 8$

$6x + 5y = 6$

(b) $4x - 2y = 8$

$2x - y = 4$

4. Solve each system by substitution.

(a) $x = 4y$

$3x - 2y = 13$

(b) $3x - 2y = 13$

$x - y = 5$

5. Solve each system.

(a) $x + 2y - z = -2$

$3x + 4y + 4z = 5$

$2x + 3y + 2z = 2$

(b) $5x - 8y + 9z = 14$

$10x - 4y - 6z = 3$

$15x + 12y - 9z = -9$

(c) $3x - 2y + z = 2$

$2x + 3y = -6$

$3x - y + z = 0$

(d) $4x - 6y + 9z = -9$

$8x - 12y + 18z = -18$

$x + 4y - 6z = 6$

(e) $x + y + z = 4$

$2x + 2y + 2z = 6$

$x + 4y - 6z = 6$

6. Solve.

(a) The sum of four times the smaller of two numbers and five times the larger is 12. The difference between three times the smaller and seven times the larger is –34. Find both numbers.

(b) The sum of three numbers is 4. The difference between three times the second number and four times the third number is –15. The difference between six times the third number and two times the first number is 14. Find all three numbers.

(c) Tickets for the July 4th parade and picnic cost \$5.00 for adults and \$3.50 for children. A total of 900 tickets were sold for \$3750. How many of each type of ticket were sold?

(d) Rhonda's coin collection consists of 331 coins with a total value of \$40.30. If the coins are nickels, dimes, and quarters, and the number of quarters and nickels is 21 more than the number of dimes, find the number of each type of coin.

(e) Manny invested a total of \$12,000 in two accounts. One account earns 6% interest annually and the other earns 71/2% interest annually. If the total interest earned from both accounts in one year is $855, find the amount invested in each account.

(f) It takes 5 hours for a boat to travel 110 miles downstream. The same boat can travel 60 miles upstream in 6 hours. What is the rate of the boat in still water and the speed of the current?

7. Find the value of each determinant.

(a) $\begin{vmatrix} 1 & 4 \\ 2 & 6 \end{vmatrix}$

(b) $\begin{vmatrix} 3 & -2 \\ 5 & 0 \end{vmatrix}$

(c) $\begin{vmatrix} 2 & 2 & 1 \\ 1 & -1 & -2 \\ 1 & 0 & 2 \end{vmatrix}$

(d) $\begin{vmatrix} -3 & 2 & -2 \\ 3 & -1 & 1 \\ 4 & -1 & 2 \end{vmatrix}$

8. Solve using Cramer's Rule.

(a) $2x + 3y = 5$

$x - 4y = 8$

(b) $2x - y = 4$

$3x + 5y = 6$

(c) $2x + 5y + 2z = 9$

$4x - 7y - 3z = 7$

$3x - 8y - 2z = 9$

$2x + 3y + 3z = -5$

$x - y - 2z = 6$

Answers

1.(a)

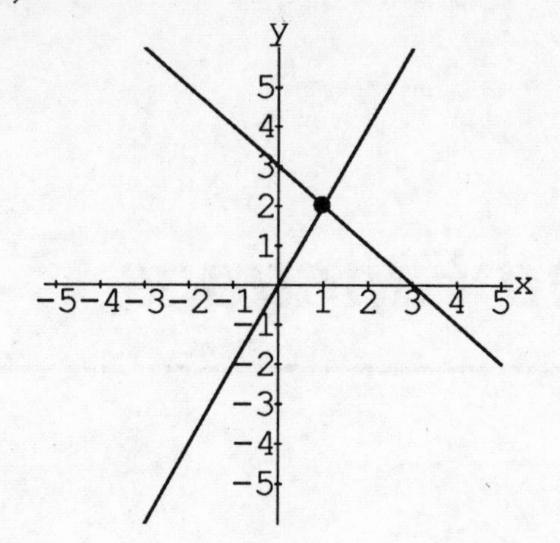

(b)

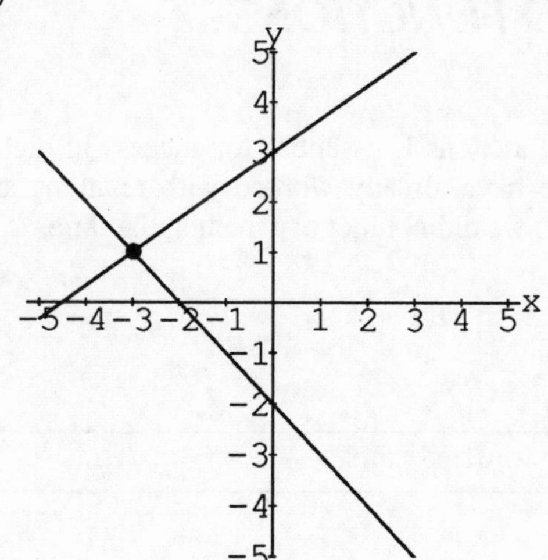

2.

(a) (4, –1)

(b) (–2, 5)

(c) (–2, –3)

(d) $(\frac{1}{2}, 3)$

3.

(a) inconsistent

(b) dependent

4.

(a) (4, 1)

(b) (3, –2)

5.(a) (3, –2, 1)

(b) $(\frac{2}{5}, -\frac{3}{4}, \frac{2}{3})$

(c) (0, –2, –2)

(d) solution is not unique

(e) ∅

6.

(a) –2 and 4

(b) 2, –1, 3

(c) 400 adults, 500 children

(d) 96 nickels, 155 dimes, 80 quarters

(e) \$9000 at $7\frac{1}{2}$%
\$3000 at 6%

(f) speed of the boat in still water is 16 mph
speed of the current is 10 mph

7.

(a) –2

(b) 10

(c) –11

(d) –3

8.

(a) (4, –1)

(b) (2, 0)

(c) (3, –1, 4)

9

Relations and Functions

9.1 RELATIONS AND FUNCTIONS

Understanding relations and functions is essential for success in higher level mathematics courses. We have already worked with relations and functions in this course, although we did not identify them at the time.

Relations

Definition. A **relation** is a set of ordered pairs.

Recall that a set is specified with brackets, and ordered pairs are written in parentheses. For example,

$$\{(1, 2), (2, 3), (2, 5), (3, 2)\}$$

is a relation. The **domain** of a relation is the set of x-coordinates, and the **range** of a relation is the set of y-coordinates.

EXAMPLE 1 Determine which of the following are relations. For the relations, identify the domain and range.

a) $\{(0, 1), (2, 3), (3, 4)\}$

b) $\{0, 1, 2, 3, 4\}$

c) $\{(x, y) | y = 2x + 1\}$

SOLUTION 1

a) $\{(0, 1), (2, 3), (3, 4)\}$ is a relation.

The domain is $\{0, 2, 3\}$.

The range is $\{1, 3, 4\}$.

b) $\{0, 1, 2, 3, 4\}$ is *not* a relation since it is not a set of ordered pairs.

c) $\{(x, y) | y = 2x + 1\}$ is read "The set of all x and y such that y equals $2x + 1$." This set does represent a set of ordered pairs, where each y-coordinate is 1 more than 2 times the x-coordinate. Some ordered pairs in this relation are $(0, 1), (1, 3), (2, 5), (-1, -1), (-2, -3)$. The domain is the set of all real numbers, since any real number can replace x. The range is the set of all real numbers since y can also be any real number.

Functions

A function is a special type of relation. In particular,

A **function** is a relation in which no x-coordinate is repeated.

For example, $\{(0, 1), (2, 3), (3, 4)\}$ is a function, while $\{(0, 1), (0, 2), (2, 3)\}$ is *not* a function.

Describing functions

Functions are described in several ways.

Type of Description	Example These are functions	Counterexample These are not functions
Set of ordered pairs	{(0, 1), (2, 3), (3, 4)}	{(0, 1), (1, 2), (1, 3)}
Table of values	x | y 0 | 1 2 | 3 3 | 4	x | y 0 | 1 1 | 2 1 | 3
Graph	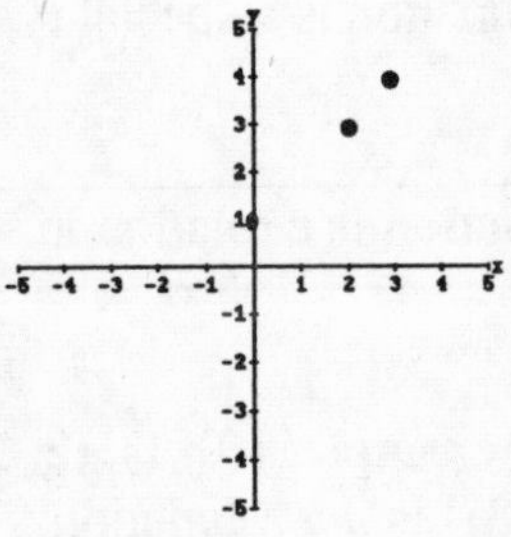	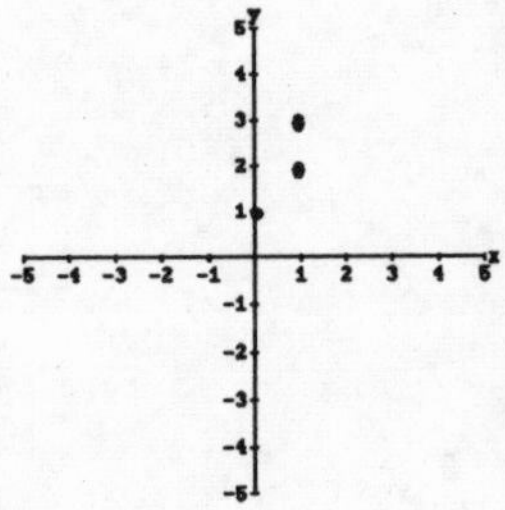
Mapping diagram	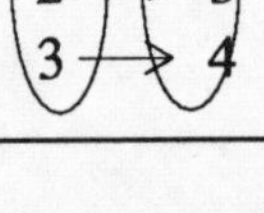 0 → 1 2 → 3 3 → 4	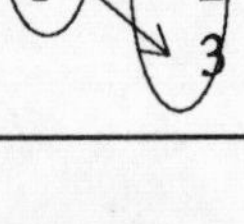 0 → 1 1 → 2 1 → 3

Notice in each counterexample that an x-coordinate is listed twice in the set of ordered pairs and the table of values. In the graph, two points line up vertically, which implied one x-coordinate matched to two y-coordinates. In the mapping diagram, two arrows coming from one x-coordinate imply that an x-coordinate is repeated and hence is not a function.

EXAMPLE 2 Identify the functions.

a) {(–1, 6), (0, 6), (1, 4)}

b)

x	y
–2	4
–1	1
0	0
1	1
2	4

c)

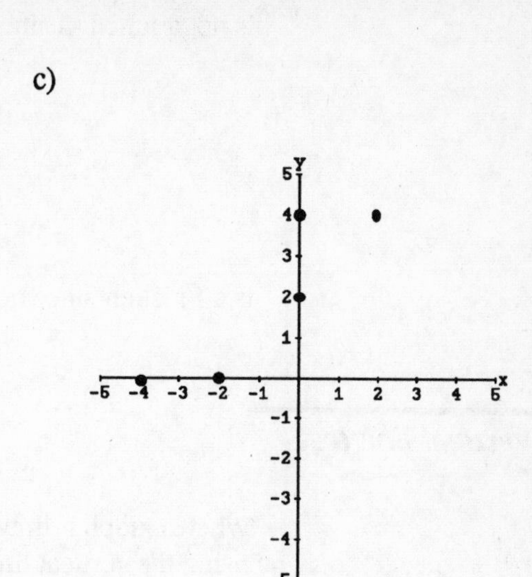

d)

SOLUTION 2

a) $\{(-1, 6), (0, 6), (1, 4)\}$ is a function since no x-coordinate is repeated.

b)

x	y
–2	4
–1	1
0	0
1	1
2	4

is a function since no x-coordinate is repeated.

c)

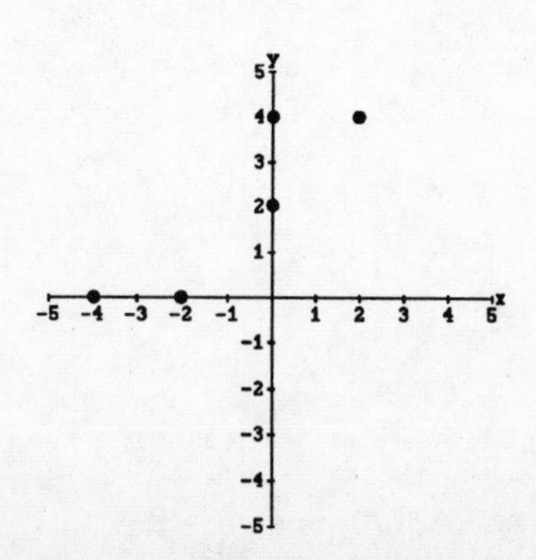

is not a function since (0, 2) and (0, 4) have the same x-coordinate.

d)

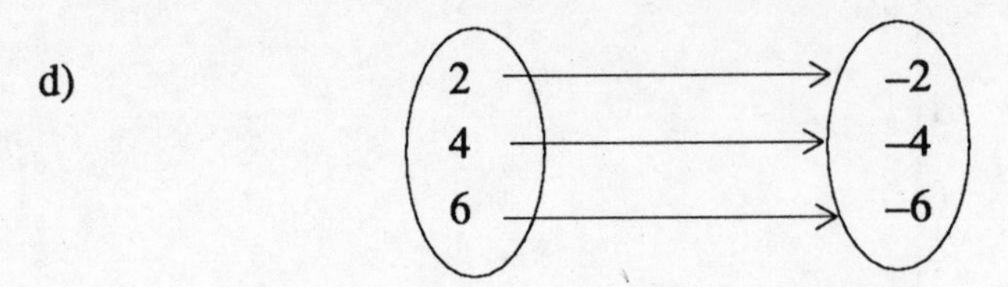

is a function since no x-coordinate is repeated.

Vertical line test

When a graph is drawn we can determine whether it represents a function by using the vertical line test.

The Vertical Line Test If a vertical line can be drawn through more than one point on a graph, that graph does *not* represent a function.

EXAMPLE 3 Identify the functions.

a) b)

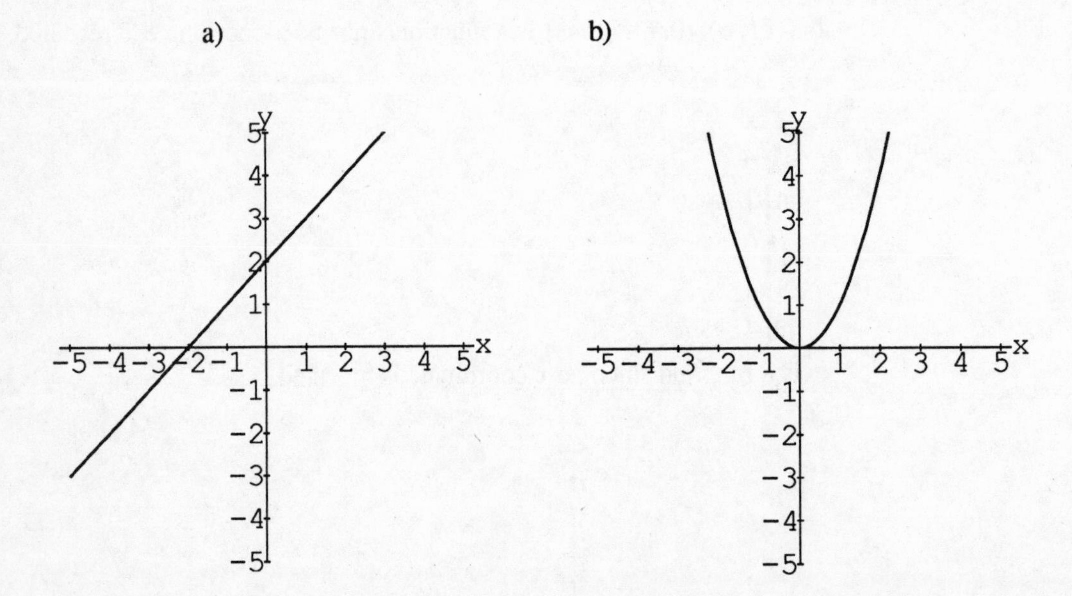

c)

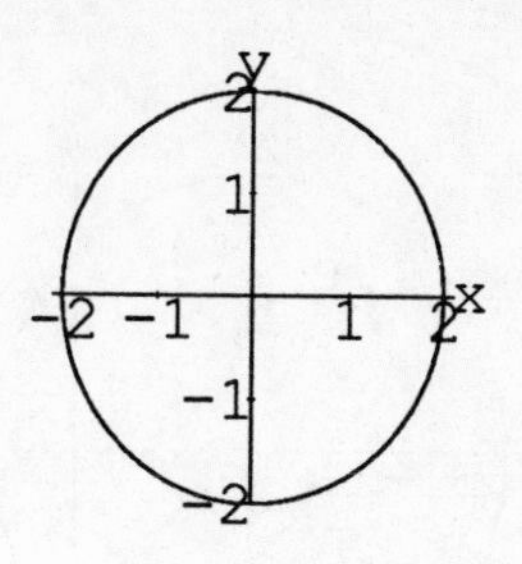

d)

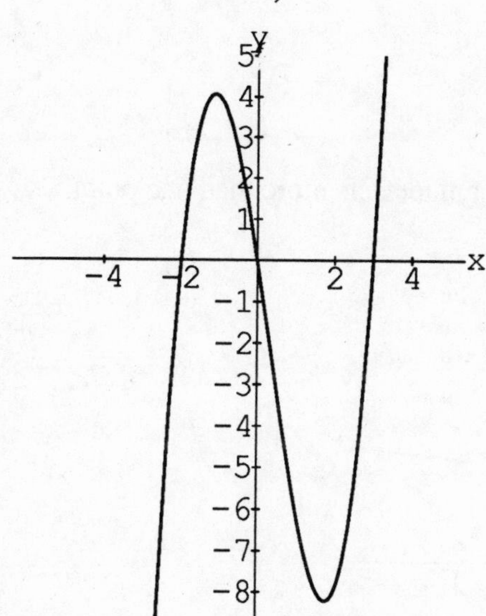

e)

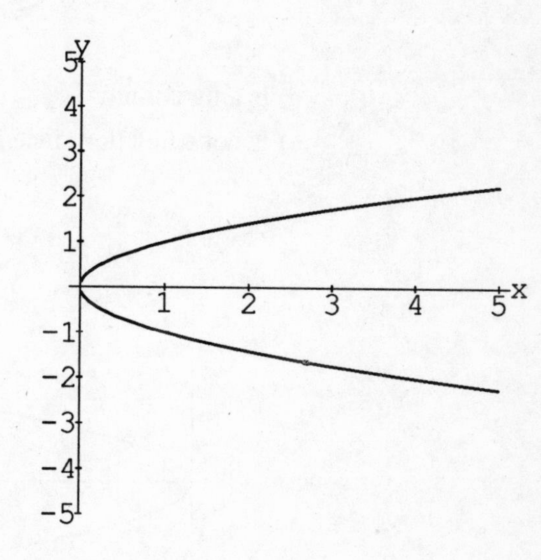

SOLUTION 3

a) Is a function.

b) Is a function.

c) Is not a function since a vertical line intersects the graph at more than one point.

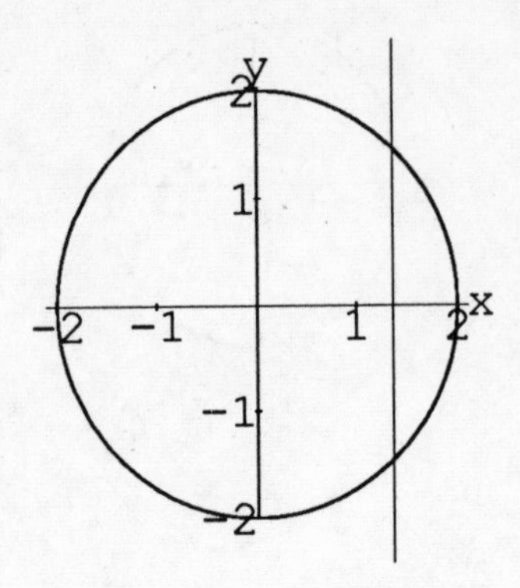

d) Is a function.

e) Is not a function since a vertical line intersects more than one point.

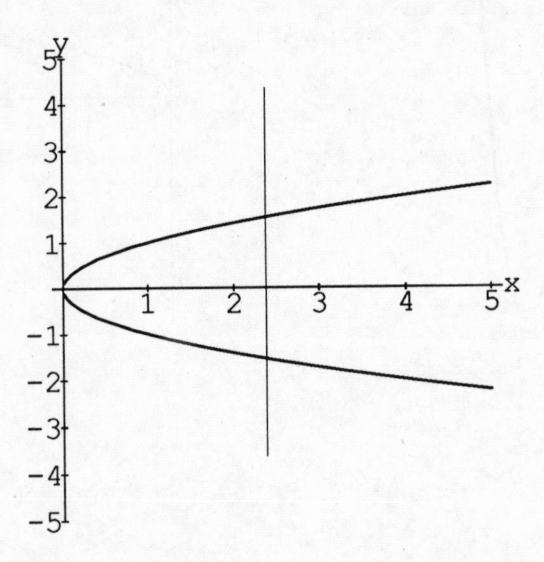

Domain and range

Now that we can determine what represents a function, we return to the concept of domain (the set of x-coordinates) and range (the set of y-coordinates). If the domain of a function is *not* stated, it is assumed to be the largest set of real numbers which can be used as x values. Since we do not allow division by 0 or negative numbers under square roots, the following steps can be used to find the domain of a function when given an equation.

1. If there are no variables in the denominator or variables under a square root, the domain is all real numbers.
2. If there is a variable in a denominator, set the denominator equal to 0 and solve. The domain is all real numbers *except* the values that make the denominator 0.
3. If there is a variable under a square root, set the radicand greater than or equal to 0, solve. The solution to the inequality is the domain.

EXAMPLE 4 Find the domain.

a) $y = 2x + 1$

b) $y = \dfrac{4}{x-2}$

c) $y = \sqrt{3x+2}$

SOLUTION 4

a) $y = 2x + 1$

The domain is the set of all real numbers since there are no variables in the denominator and no square roots.

b) $y = \dfrac{4}{x-2}$

$x - 2 = 0$ — Set the denominator equal to 0.

$x = 2$ — Solve by adding 2 to both sides.

The domain is the set of all real numbers except 2, written $\{x \mid x \neq 2\}$.

c) $y = \sqrt{3x+2}$

$3x + 2 \geq 0$ — Set the radicand greater than or equal to 0.

$3x \geq -2$ — Subtract 2 from both sides.

$x \geq -\dfrac{2}{3}$ — Divide by 3.

The domain is the set of all real numbers greater than or equal to $-\dfrac{2}{3}$,

written $\{x \mid x \geq -\dfrac{2}{3}\}$.

Domain and range given a graph

To find the domain and range when looking at a graph, look for the largest and smallest values to the left and right for the domain, and up and down for the range.

EXAMPLE 5 State the domain and range.

a)

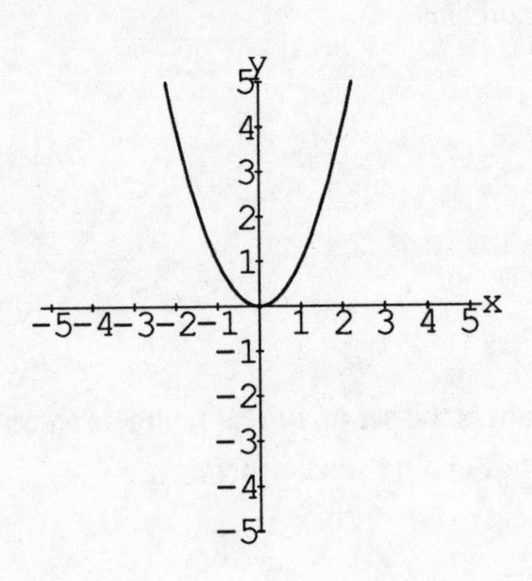

b)

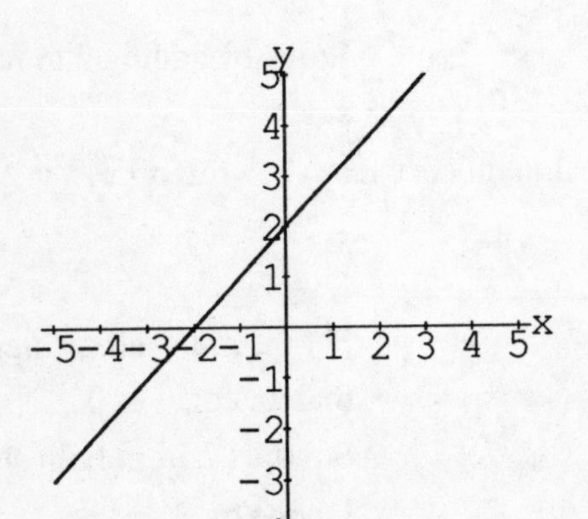

c)

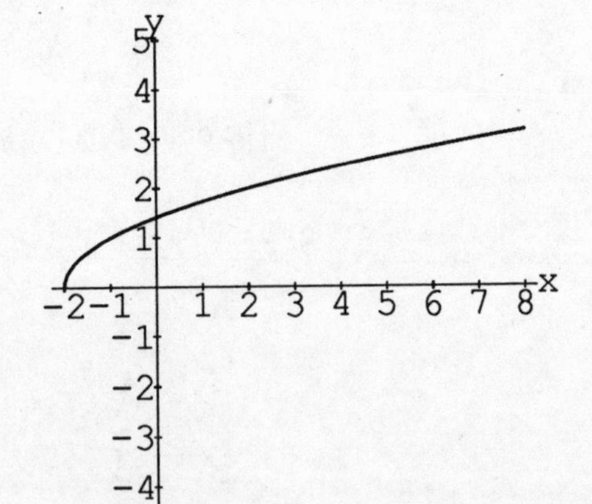

SOLUTION 5

a) The domain is the set of all real numbers since the arrows indicate that the graph will continue. The range is $\{y \mid y \geq 0\}$ since the smallest y value is 0 and the graph continues upward.

b) The domain is the set of all real numbers and the range is the set of all real numbers. The arrows indicate that the line continues.

c) The domain is the set of all real numbers greater than or equal to –2, written $\{x \mid x \geq -2\}$. The range is the set of all real numbers greater than or equal to 0, written $\{y \mid y \geq 0\}$.

9.2 FUNCTION NOTATION

f(x) notation

In the past we have discussed the line $y = 2x + 1$, which we now know is a function. We use special notation for functions so that

$y = 2x + 1$	in function notation is	$f(x) = 2x + 1$
$y = x^2$	in function notation is	$g(x) = x^2$
$y = \sqrt{x}$	in function notation is	$h(x) = \sqrt{x}$

$f(x)$ is read "f of x", $g(x)$ is read "g of x", and $h(x)$ is read "h of x."

Finding f(x)

Consider the function $f(x) = 2x + 1$. " Find $f(3)$" means find the value of the function (the y-coordinate) when $x = 3$. You are being asked to substitute 3 for x and simplify the result:

$$f(3) = 2(3) + 1$$
$$= 6 + 1$$
$$= 7$$
$$\text{so } f(3) = 7.$$

EXAMPLE 6 If $f(x) = 2x + 1$, find

a) $f(0)$

b) $f(-4)$

c) $f(a)$

d) $f(*)$

e) $f(x + h)$

SOLUTION 6

a) $f(0) = 2(0) + 1$ — Substitute $x = 0$.

$= 0 + 1$ — Simplify.

$= 1$

b) $f(-4) = 2(-4) + 1$ — Substitute $x = -4$.

$= -8 + 1$ — Simplify.

$= -7$

c) $f(a) = 2(a) + 1$ — Substitute $x = a$.

$= 2a + 1$ — Simplify. Stop here because $2a$ and 1 are not similar terms.

d) $f(*) = 2(*) + 1$ — Substitute $x = *$.

$= 2* + 1$ — Simplify.

e) $f(x + h) = 2(x + h) + 1$ — Substitute $x = x + h$.

$= 2x + 2h + 1$ — Use the distributive property.

EXAMPLE 7 If $g(x) = 3x^2 + 2x - 1$ find

a) $g(0)$

b) $g(-2)$

c) $g(*)$

d) $g(x + h)$

SOLUTION 7

a) $g(0) = 3(0)^2 + 2(0) - 1$ — Substitute $x = 0$.

$= 0 + 0 - 1$ — Simplify.

$= -1$

b) $g(-2) = 3(-2)^2 + 2(-2) - 1$ — Substitute $x = -2$.

$= 3(4) - 4 - 1$ — Simplify exponents before multiplying.

$= 12 - 4 - 1$ — Multiply.

$= 7$

c) $g(*) \qquad = 3(*)^2 + 2(*) - 1$ — Substitute $x = *$.

d) $g(x + h)$

$= 3(x + h)^2 + 2(x + h) - 1$ — Substitute $x = x + h$.

$= 3(x^2 + 2xh + h^2) + 2(x + h) - 1$ — $(x + h)^2 = (x + a)(x + h)$ $= x^2 + 2xh + h^2$.

$= 3x^2 + 6xh + 3h^2 + 2x + 2h - 1$ — Use the distributive property.

Operations with functions

Just as we can add, subtract, multiply and divide numbers, we can add, subtract, multiply and divide functions.

Definitions. If f and g are functions,

$(f + g)(x) = f(x) + g(x)$

$(f - g)(x) = f(x) - g(x)$

$(fg)(x) = f(x)g(x)$

$\frac{f}{g}(x) = \frac{f(x)}{g(x)}, \; g(x) \neq 0$

EXAMPLE 8 If $f(x) = 2x + 5$ and $g(x) = x^2 - 2x + 1$, find

a) $(f + g)(3)$

b) $(fg)(-1)$

c) $\frac{f}{g}(0)$

SOLUTION 8

a) $(f+g)(3)$

$= f(3) + g(3)$	Use $(f+g)(x) = f(x) + g(x)$.
$= 2(3) + 5 + (3)^2 - 2(3) + 1$	Substitute $x = 3$.
$= 6 + 5 + 9 - 6 + 1$	Simplify $x = 3$.
$= 15$	

b) $(fg)(-1)$

$= f(-1)g(-1)$	Use $(fg)(x) = f(x)g(x)$.
$= [2(-1) + 5][(-1)^2 - 2(-1) + 1]$	Substitute $x = -1$.
$= (3)(4)$	Simplify inside each set of brackets.
$= 12$	Multiply.

c) $\frac{f}{g}(0)$

$= \frac{f(0)}{g(0)}$	Use $\frac{f}{g}(x) = \frac{f(x)}{g(x)}$.
$= \frac{2(0)+1}{(0)^2 - 2(0) + 1}$	Substitute $x = 0$.
$= \frac{1}{1}$	Simplify the numerator and denominator.
$= 1$	

EXAMPLE 9 If $f(x) = 3x^2 - 5x - 4$ and $g(x) = 2x + 3$, find

a) $(f-g)(x)$

b) $(fg)(x)$

c) $f(f,g)(x)$

SOLUTION 9

a) $(f-g)(x)$

$= f(x) - g(x)$	Use the definition of function subtraction.
$= (3x^2 - 5x - 4) - (2x + 3)$	Substitute.
$= 3x^2 - 5x - 4 - 2x - 3$	Subtract.

$= 3x^2 - 7x - 7$ Combine similar terms.

b) $(fg)(x)$

$= f(x)g(x)$ Use the definition of function multiplication.

$= (3x^2 - 5x - 4)(2x + 3)$ Substitute.

$= 6x^3 + 9x^2 - 10x^2 - 15x - 8x - 12$ Distribute.

$= 6x^3 - x^2 - 23x - 12$ Combine similar terms.

c) $\frac{f}{g}(x)$

$= \frac{f(x)}{g(x)}$ Use the definition of function division.

$= \frac{3x^2 - 5x - 4}{2x + 3}$ Substitute.

Composition of functions

In addition to the four basic operations (addition, subtraction, multiplication, division), we can also compose functions.

Definition. The composition of functions f and g, written $f o g$ is $f o g = f(g(x))$

The composition of functions indicates an order of evaluating the functions. $(f o g)(3) = f(g(3))$ means first evaluate $g(3)$, then evaluate f at that value.

EXAMPLE 10 If $f(x) = 3x + 1$ and $g(x) = -2x - 4$, find $(f o g)(3)$.

SOLUTION 10

$(f o g)(3) = f(g(3))$ Use the definition of composition of functions.

$= f(-2(3) - 4)$ Find $g(3)$.

$= f(-10)$ Now evaluate $f(-10)$.

$= 3(-10) + 1$ — Substitute $x = -10$.

$= -31$

EXAMPLE 11 If $f(x) = 3x + 1$ and $g(x) = -2x - 4$, find $(gof)(3)$.

SOLUTION 11

$(gof)(3) = g(f(3))$

$= g(3(3) + 1)$ — Find $f(3)$.

$= g(10)$ — Now evaluate $g(10)$.

$= -2(10) - 4$ — Substitute $x = 10$.

$= -24$

Notice from Examples 10 and 11 that in general $gof \neq fog$.

9.3 TYPES OF FUNCTIONS

In this section we will identify several types of functions. This is not a comprehensive list and if you continue to study mathematics you will probably encounter several other types of functions. The following table presents the name and form of the functions and gives an example and graph of each type.

Name	Form	Example	Graph
Linear function	$f(x) = ax + b$	$f(x) = -2x + 3$	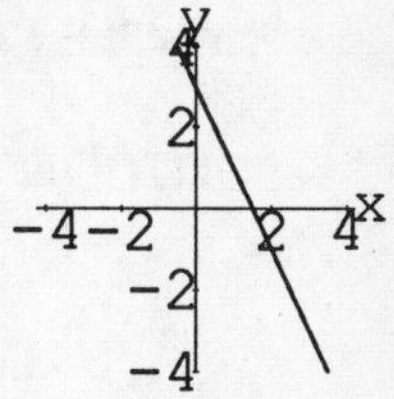
Constant function	$f(x) = a$	$f(x) = 3$	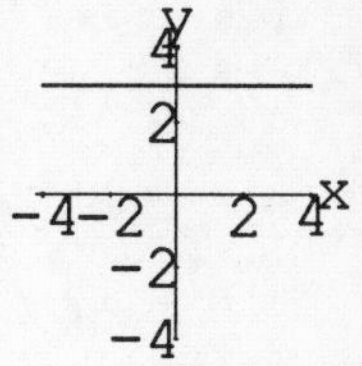
Quadratic function	$f(x) = ax^2 + bx + c$	$f(x) = x^2 - 2x - 1$	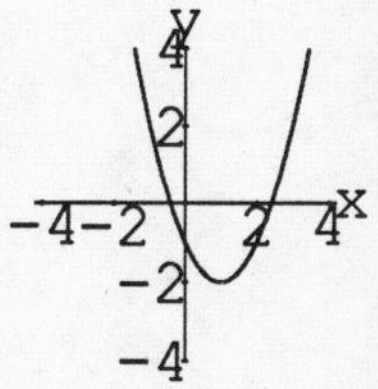
Polynomial function	$f(x) = a_n x^n + a_{n-1} x^{n-1} + \ldots + a_1 x + a_0$	$f(x) = x^3 + x^2 - 2x$	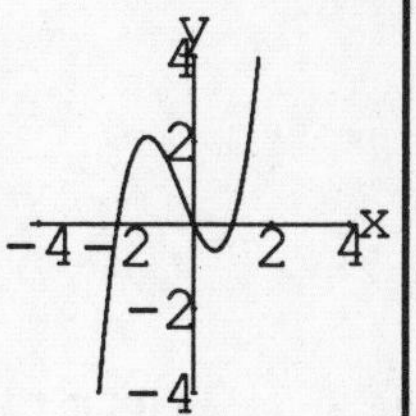
Square root function	$f(x) = \sqrt{ax + b}$	$f(x) = \sqrt{x}$	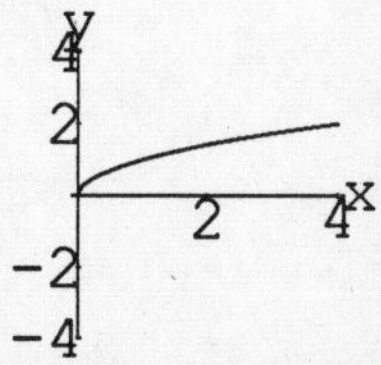

Each of the graphs in the table can be obtained using a table of values or in the case of linear functions, by methods previously presented. The following examples will provide details for each type of function.

EXAMPLE 12 Graph $f(x) = \frac{2}{3}x - 4$ and identify the type of function.

SOLUTION 12

Replacing $f(x)$ with y gives us the equation

$$y = \frac{2}{3}x - 4$$

which represents the equation of a line written in slope intercept form. The slope m equals $\frac{2}{3}$ and the y-intercept equals –4. Begin at the y-intercept and then rise 2 and run 3:

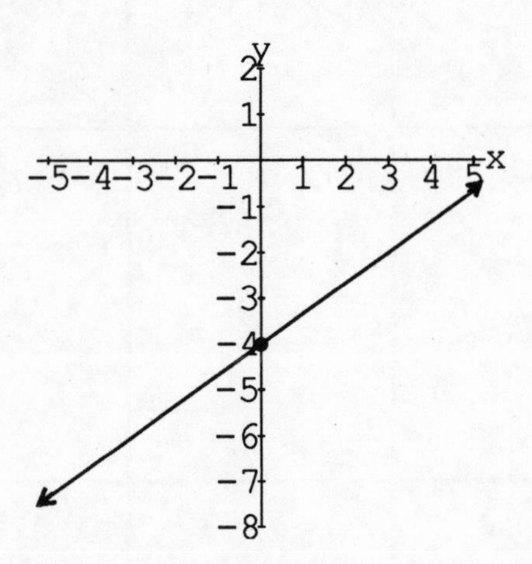

$f(x) = \frac{2}{3}x - 4$ is a linear function. Note that the graph of every linear function is a line.

EXAMPLE 13 Graph $f(x) = -2$ and identify the type of function.

SOLUTION 13

Replacing $f(x)$ with y gives us the equation

$y = -2$

which represents the equation of a horizontal line passing through the point (0, –2):

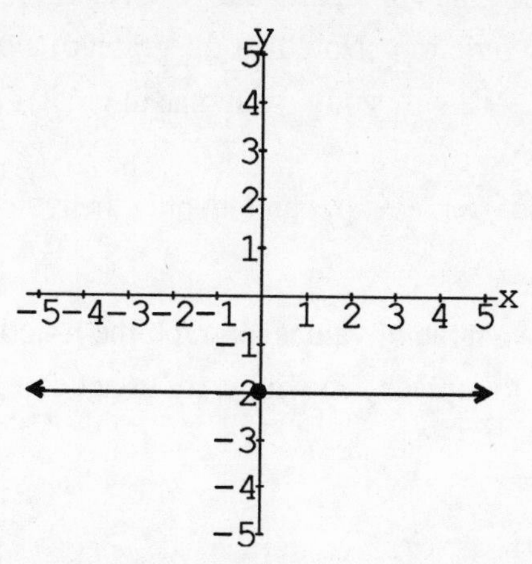

$f(x) = -2$ is a constant function. Note that the graph of every constant function is a horizontal line.

EXAMPLE 14 Graph $f(x) = x^2 - 4$ and identify the type of function.

SOLUTION 14

Replacing $f(x)$ with y gives us the equation

$y = x^2 - 4$

We can use a table of values to graph this function.

x	y
-2	0
-1	-3
0	-4
1	-3
2	0

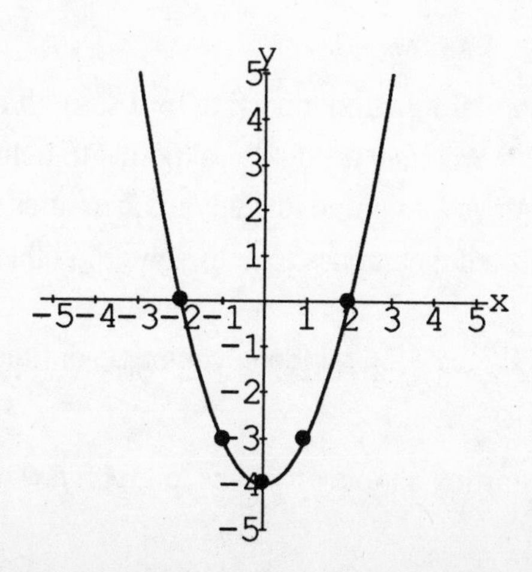

This is the graph of a parabola. The function $f(x) = x^2 - 4 = x^2 + 0x - 4$ is a quadratic function. Note that the graph of every quadratic function is a parabola. We will study parabolas in greater detail in Chapter 10.

EXAMPLE 15 Graph $f(x) = x^3 - x^2 - 6x$ and identify the type of function.

SOLUTION 15

We'll use a table of values to graph the function. Although you can continue to replace $f(x)$ with y, try to get comfortable using $f(x)$.

x	$f(x)$
-3	−18
-2	0
-1	4
0	0
1	−6
2	−8
3	0

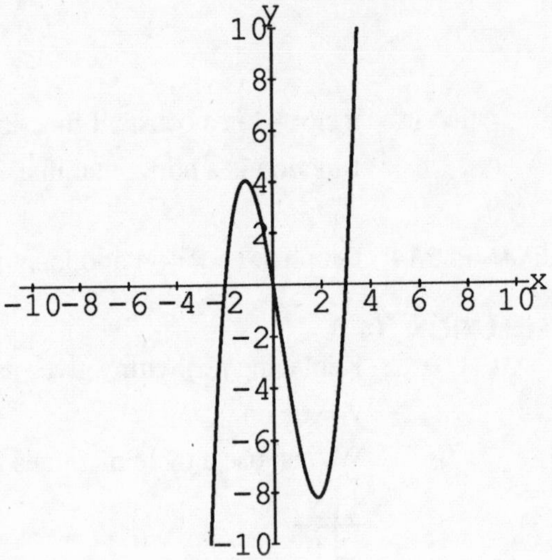

A polynomial function graphs to be a smooth curve that may have several "humps." You may need several points to help determine the shape of the graph. As you continue to study mathematics you may use more sophisticated techniques to help you graph these polynomial functions.

EXAMPLE 16 Graph $f(x) = \sqrt{x+2}$ and identify the type of function.

SOLUTION 16

Before starting a table of values to graph $f(x) = \sqrt{x+2}$, let's find the domain:

$x + 2 \geq 0$ — Set the radicand greater than or equal to 0.

$x \geq -2$ — Solve for x.

Since the domain consists of x values greater than or equal to –2, we create the following table choosing x and finding $f(x)$.

x	$f(x)$
–2	0
–1	1
2	2
7	3

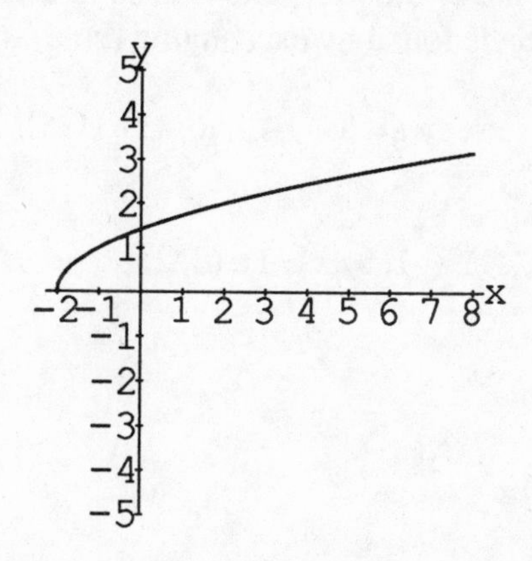

The function $f(x) = \sqrt{x+2}$ is a square root function. Note that the graph of every square root function will be half of a parabola.

9.4 INVERSE FUNCTIONS

Recall the mapping diagram of a function in section 9.1

f: 0 → 1, 2 → 3, 3 → 4

$f(0) = 1$
$f(2) = 3$
$f(3) = 4$

In this section we will study the inverse function, f^{-1}, read "f inverse," which maps elements from the range back to elements in the domain:

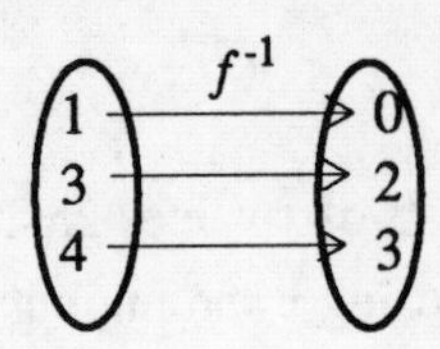

$f^{-1}(1) = 0$
$f^{-1}(3) = 2$
$f^{-1}(4) = 3$

Finding inverses with ordered pairs

When a list of ordered pairs is used to describe a function, the inverse of that function is found by exchanging the x- and y-coordinates.

EXAMPLE 17 Find the inverse of $f = \{(-1, -3), (0, -1), (1, 1), (2, 3)\}$.

SOLUTION 17

$$f^{-1} = \{(-3, -1), (-1, 0), (1, 1), (3, 2)\}$$

Finding inverses given an equation

We can use the idea of exchanging x and y to find the inverse of a function that is given as an equation.

To find f^{-1}:
1. Rewrite $f(x)$ as y.
2. Exchange x and y.
3. Solve for y.
4. Rewrite y as $f^{-1}(x)$.

EXAMPLE 18 Find the inverse of each function.

a) $f(x) = 2x - 1$

b) $f(x) = x^3 - 2$

SOLUTION 18

a) $f(x) = 2x - 1$

$y = 2x - 1$	Rewrite $f(x)$ as y.
$x = 2y - 1$	Exchange x and y.
$x + 1 = 2y$	Add 1 to both sides of the equation.
$\frac{x+1}{2} = y$	Divide by 2.
$\frac{x+1}{2} = f^{-1}(x)$	Rewrite y as $f^{-1}(x)$.

b) $f(x) = x^3 - 2$

$y = x^3 - 2$	Rewrite $f(x)$ as y.
$x = y^3 - 2$	Exchange x and y.
$x + 2 = y^3$	Add 2 to both sides of the equation.
$\sqrt[3]{x+2} = \sqrt[3]{y^3}$	Take the cube root of both sides.
$\sqrt[3]{x+2} = y$	Simplify $\sqrt[3]{y^3} = y$.
$\sqrt[3]{x+2} = f^{-1}(x)$	Rewrite y as $f^{-1}(x)$.

One-to-one functions

If we want the inverse of a function to be a function, we must be careful that the original function is one-to-one.

> **Definition**. Each x-coordinate in a one-to-one function corresponds to exactly one y-coordinate.

Given a set of ordered pairs, no two ordered pairs have the same y-coordinate in a one-to-one function. On the graph of a one-to-one function, no horizontal line intersects the graph in more than one point.

Consider the graph of $f(x) = x^2$:

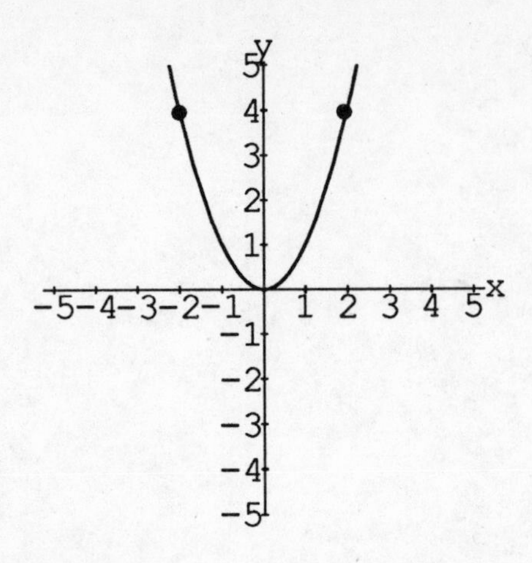

The ordered pairs (–2, 4) and (2, 4) are on the graph of $f(x) = x^2$, which means the ordered pairs (4, –2) and (4, 2) would be on the inverse of $f(x) = x^2$. But, a repeated x-coordinate implies that this inverse would *not* be a function.

Only one-to-one functions have inverses that are functions.

EXAMPLE 19 For each one-to-one function, find the inverse.

a) {(2, 3), (4, 5), (–1, 5)}

b) {(–2, –4), (–1, –1), (0, 0), (1, –1), (2, –4)}

c) {(0, 0), (1, 1), (4, 2)}

d) $f(x) = \frac{1}{3}x - 1$

e) $f(x) = x^2 - 1$

SOLUTION 19

a) Not a one-to-one function since two ordered pairs have the same y-coordinate.

b) Not a one-to-one function since y-coordinates are repeated.

c) This is a one-to-one function. The inverse is {(0, 0), (1, 1), (2, 4)}.

d) $f(x) = \frac{1}{3}x - 1$ is a linear function whose graph is

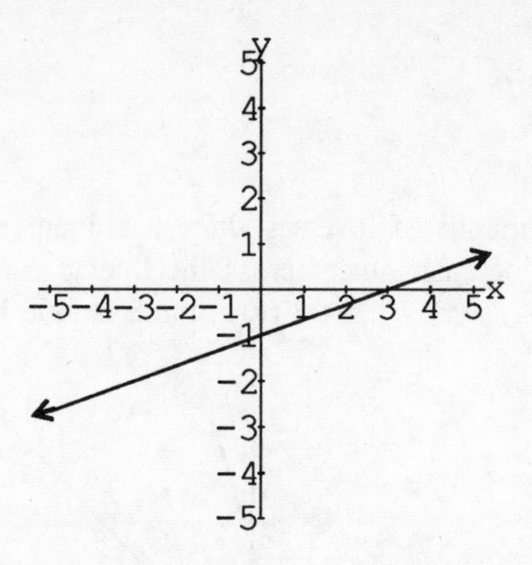

This graph passes the horizontal line test (no horizontal line will intersect the graph in more than one point) and is a one-to-one function.

$y = \frac{1}{3}x - 1$	Rewrite $f(x)$ as y.
$x = \frac{1}{3}y - 1$	Exchange x and y.
$x + 1 = \frac{1}{3}y$	Add 1 to both sides.
$3x + 3 = y$	Multiply both sides by 3.
$3\ x + 3 = f^{-1}(x)$	Rewrite y as $f^{-1}(x)$.

e) $f(x) = x^2 - 1$ is a quadratic function. The graph is a parabola and does *not* pass the horizontal line test:

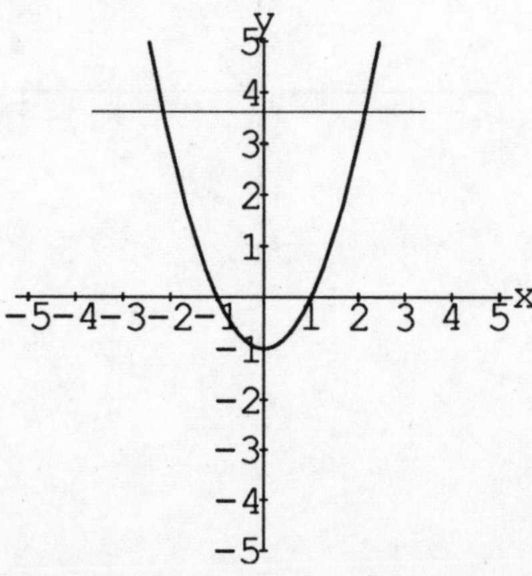

The function $f(x) = x^2 - 1$ is *not* a one-to-one function.

The graphing relationship of f and f^{-1}

The graphs of inverses share a unique relationship. They are mirror images of each other across the line $y = x$. If we graph $f(x) = \frac{1}{3}x - 1$ and $f^{-1}(x) = 3x + 3$ (found in Example 19 d), we have

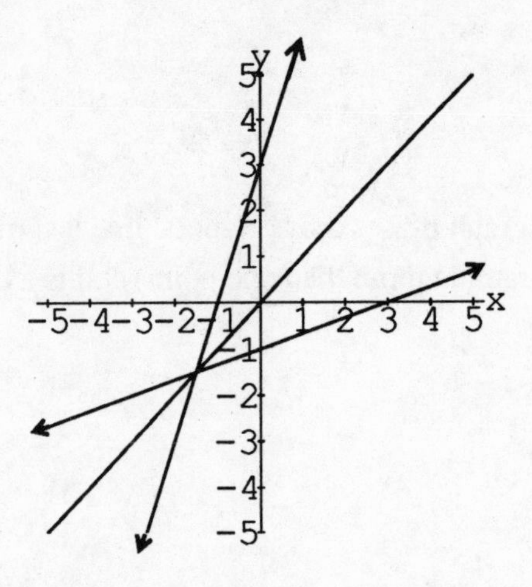

EXAMPLE 20 For each one-to-one function, graph the inverse with a dashed curve.

a)

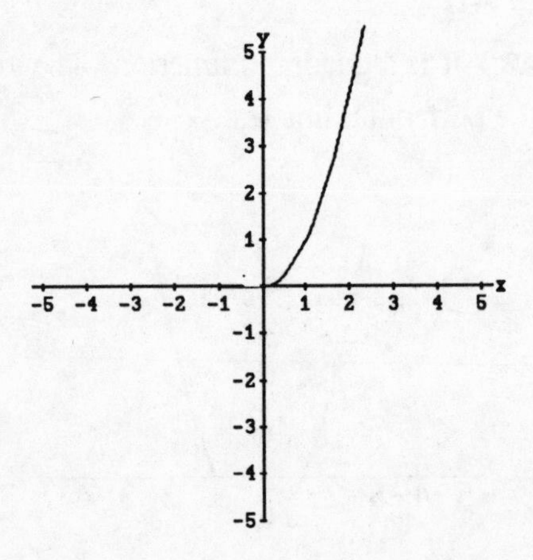

b)

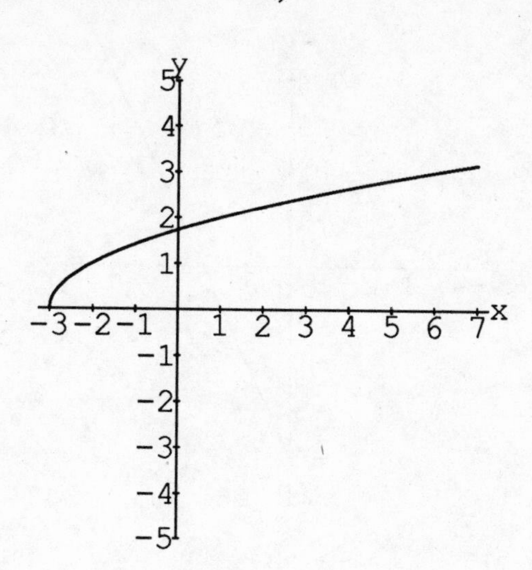

c)

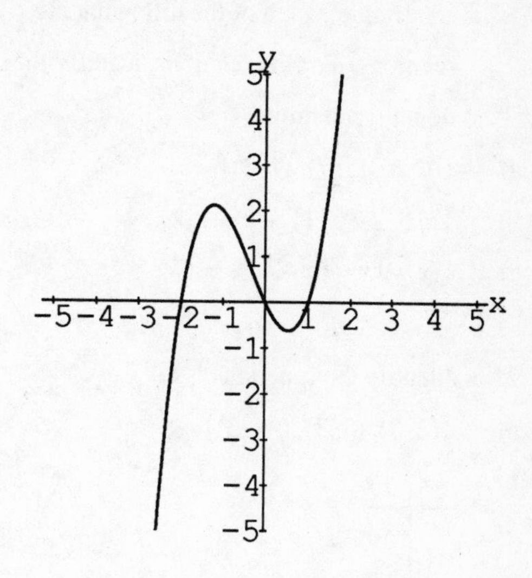

SOLUTION 20

a)

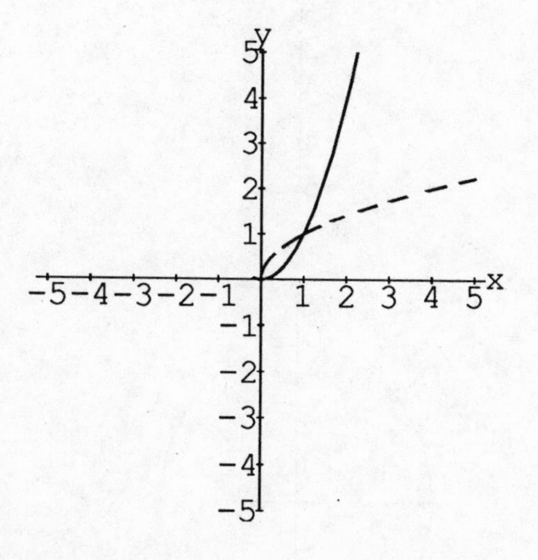

b)Note that (–3, 0) is the starting point of f, so (0, –3) must be on f^{-1} .

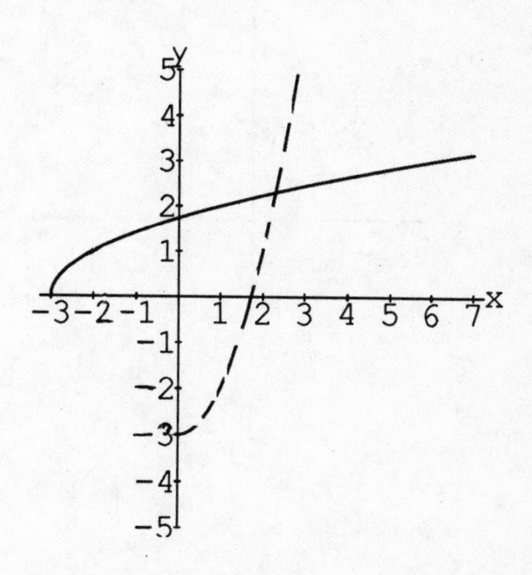

c) This is not a one-to-one function.

Practice Exercises

1. Determine which of the following are relations. For the relations, identify the domain and range.

a) $\{(2, 5), (5, 7), (7, 3)\}$

b) $\{3, 5, 7, 9, 11\}$

c) $\{(x, y) \mid y = 6x - 4\}$

2. Identify the functions.

a) $\{(2, 7), (0, 3), (-4, 3)\}$

b)

x	y
–2	8
–1	1
0	0
1	1
2	8

c)

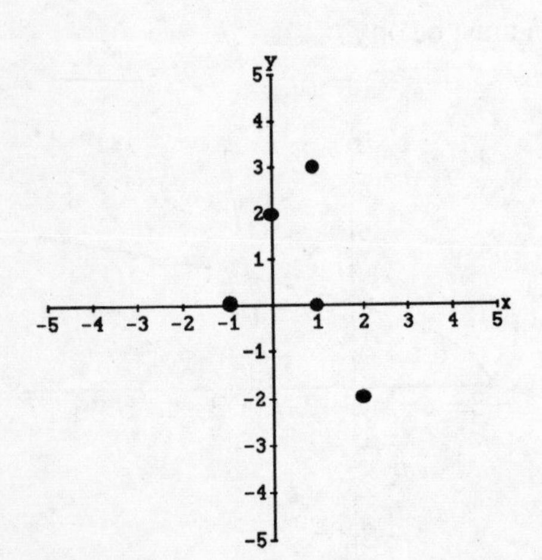

d)

–1	1
3	–3
–5	5

3. Identify the functions.

a)

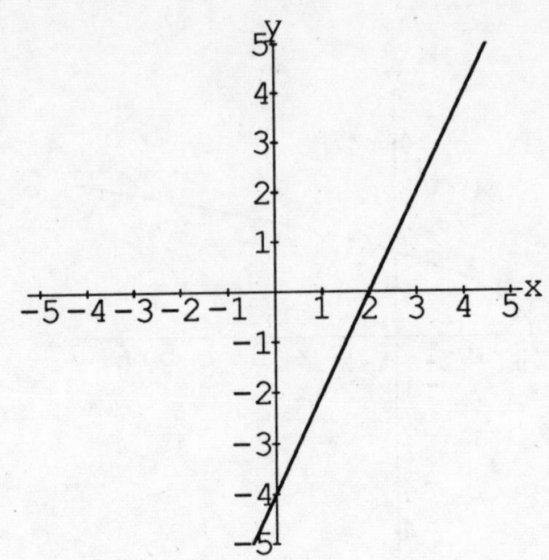

b)

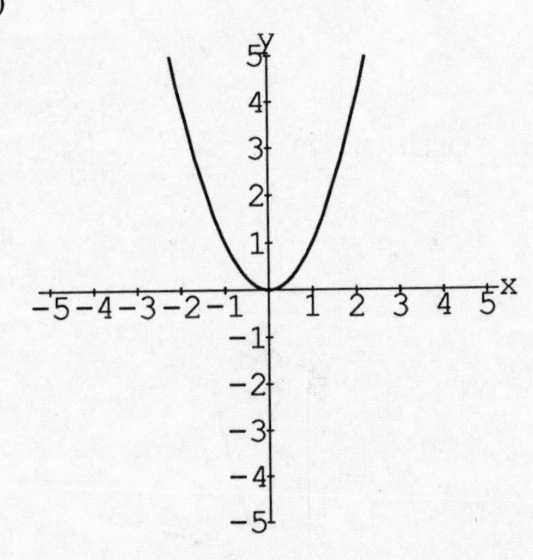

c)

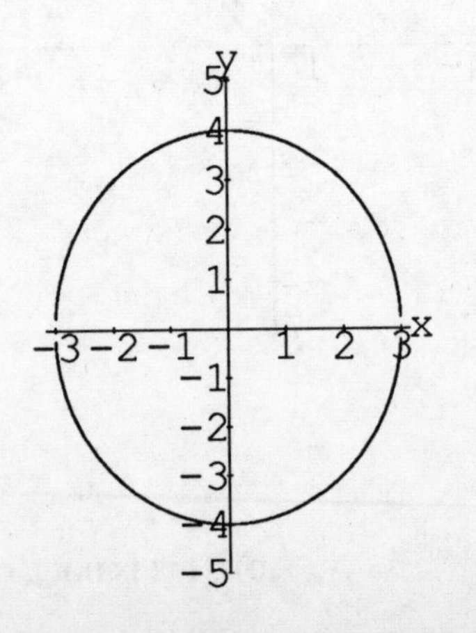

d)

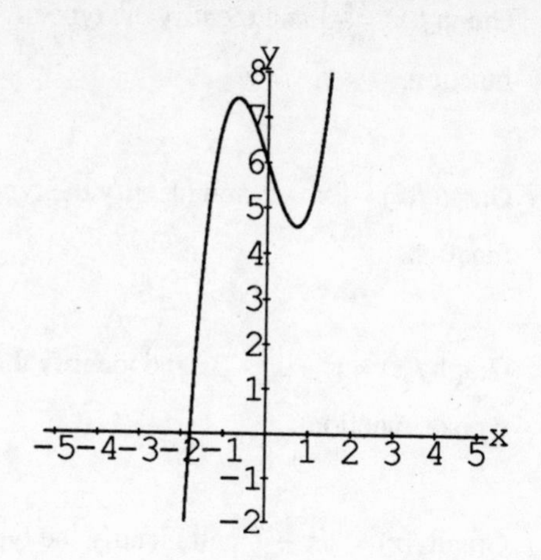

e)

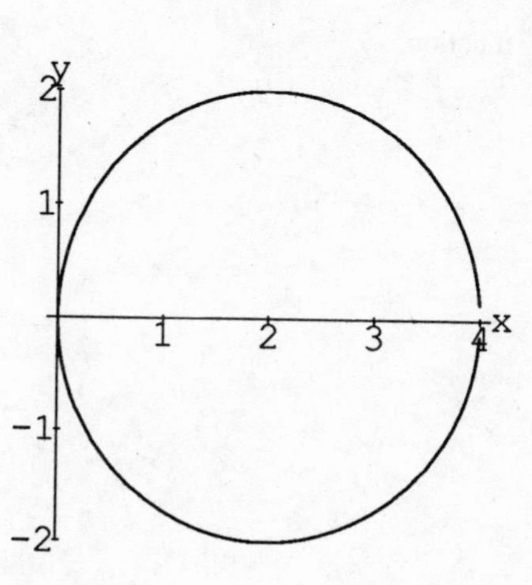

4. Find the domain.

a) $y = 3x + 4$

b) $y = \dfrac{6}{2 + x}$

c) $y = \sqrt{x - 8}$

5. State the domain and range.

a)

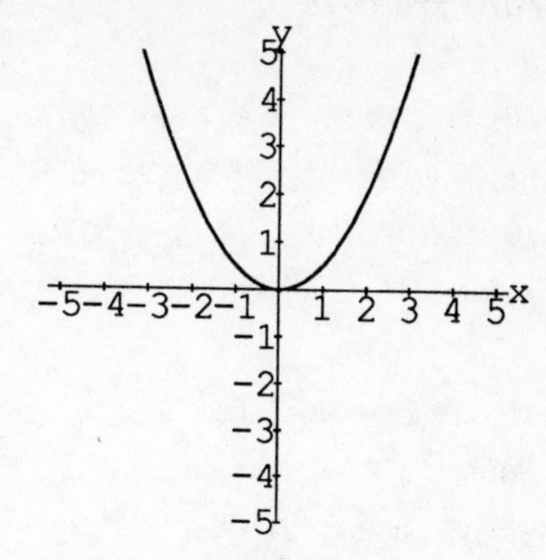

b)

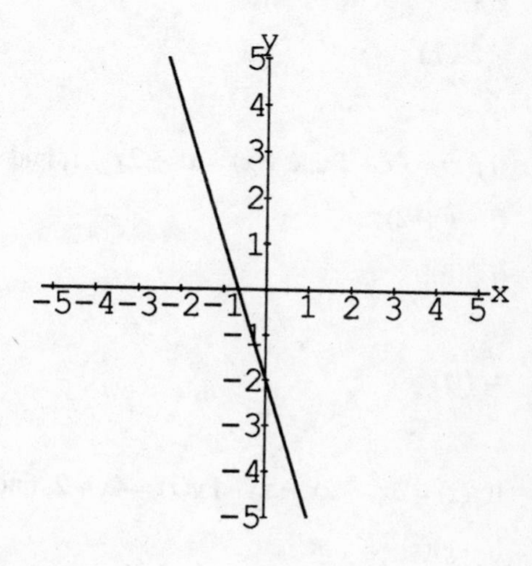

c)

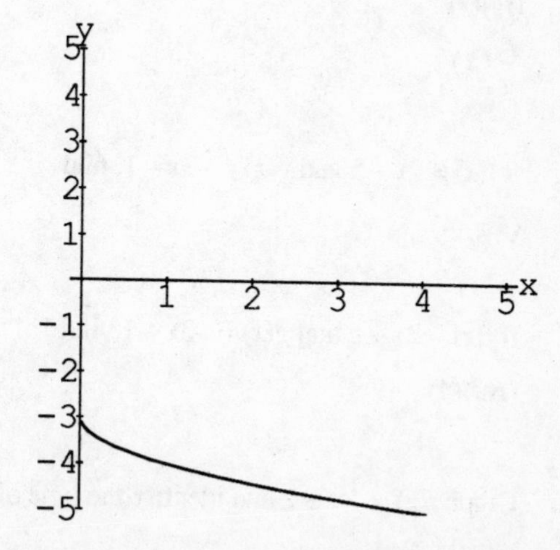

6. If $f(x) = 4x - 2$, find

a) $f(0)$

b) $f(3)$

c) $f(b)$

d) $f(\#)$

e) $f(x - h)$

7. If $g(x) = 2x^2 + 3x - 1$, find

a) $g(0)$

b) $g(-3)$

c) $g(\#)$

d) $g(x - h)$

8. If $f(x) = 4x + 1$ and $g(x) = x^2 + 2x - 1$, find

a) $(f + g)(-2)$

b) $(fg)(3)$

c) $\frac{f}{g}(0)$

9. If $f(x) = 3x^2 - 2x + 5$ and $g(x) = 4x - 2$, find

a) $(f - g)(x)$

b) $(fg)(x)$

c) $\frac{f}{g}(x)$

10. If $f(x) = 2x + 5$ and $g(x) = -3x - 1$, find $(f o g)(4)$.

11. If $f(x) = 2x + 5$ and $g(x) = -3x - 1$, find $(g o f)(4)$.

12. Graph $f(x) = \frac{3}{5}x + 2$ and identify the type of function.

13. Graph $f(x) = -3$ and identify the type of function.

14. Graph $f(x) = 2x^2 - 1$ and identify the type of function.

15. Graph $f(x) = x^3 + x^2 - 3x$ and identify the type of function.

16. Graph $f(x) = \sqrt{x - 2}$ and identify the type of function.

Answers

1.

a) $\{(2, 5), (5, 7), (7, 3)\}$ is a relation.

The domain is $\{2, 5, 7\}$.

The range is $\{3, 5, 7\}$.

b) $\{3, 5, 7, 9, 11\}$ is not a relation since it is not a set of ordered pairs.

c) $\{(x, y) \mid y = 6x - 4\}$ is a relation.

The domain is the set of all real numbers.

The range is the set of all real numbers.

2.

a) Is a function.

b) Is a function.

c) Is not a function.

d) Is a function.

3.

a) Is a function.

b) Is a function.

c) Is not a function.

d) Is a function.

e) Is not a function.

4.

a) $\{x \mid x$ is an element of R$\}$

b) $\{x \mid x \neq -2\}$.

c) $\{x \mid x \geq 8\}$

5.

a) $\{x \mid x$ is an element of R$\}$

$\{y \mid y \geq 0\}$

b) $\{x \mid x$ is an element of R$\}$

$\{y \mid y \geq 0\}$

c) $\{x \mid x \leq -3\}$

$\{y \mid y \leq 0\}$

6.

a) $f(0) = -2$

b) $f(3) = 10$

c) $f(b) = 4b - 2$

d) $f(\#) = 4\# - 2$

e) $f(x - h) = 4x - 4h - 2$

7.

a) $g(0) = -1$

b) $g(-3) = 8$

c) $g(\#) = 2(\#)^2 + 3(\#) - 1$

d) $g(x - h) = 2x^2 - 4xh + 2h^2 + 3x - 3h - 1$

8

a) $(f + g)(-2) = -8$

b) $(fg)(3) = 182$

c) $\dfrac{f}{g}(0) = -1$

9.

a) $(f - g)(x) = 3x^2 - 6x + 7$

b) $(fg)(x) = 12x^3 - 14x^2 + 24x - 10$

c) $\dfrac{f}{g}(x) = \dfrac{3x^2 - 2x + 5}{4x - 2}$

10. $(fog)(4) = -21$

11. $(gof)(4) = -27$

12.

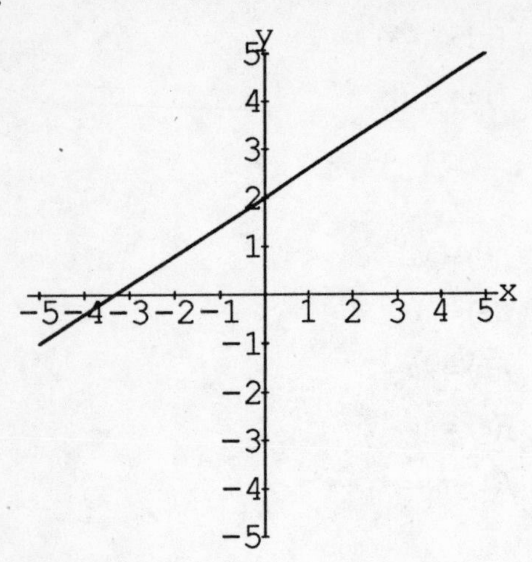

linear function

13.

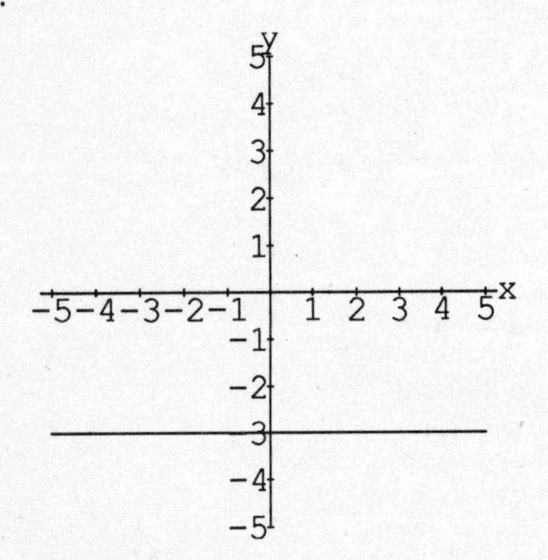

constant function

14.

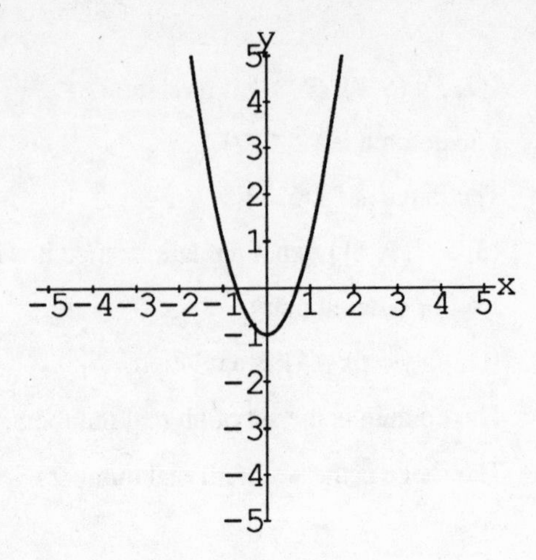

quadratic function

15.

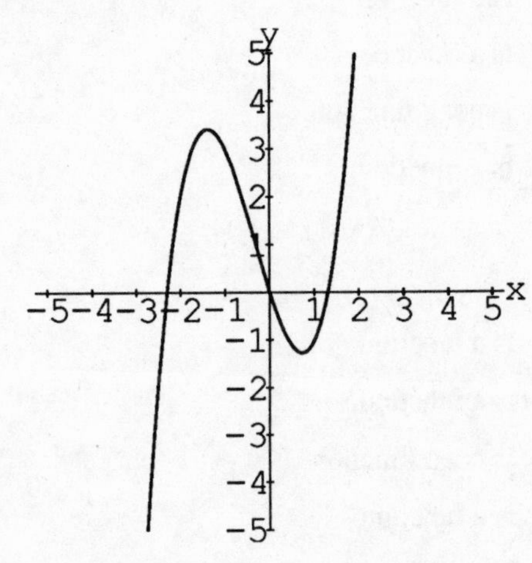

polynomial function

16.

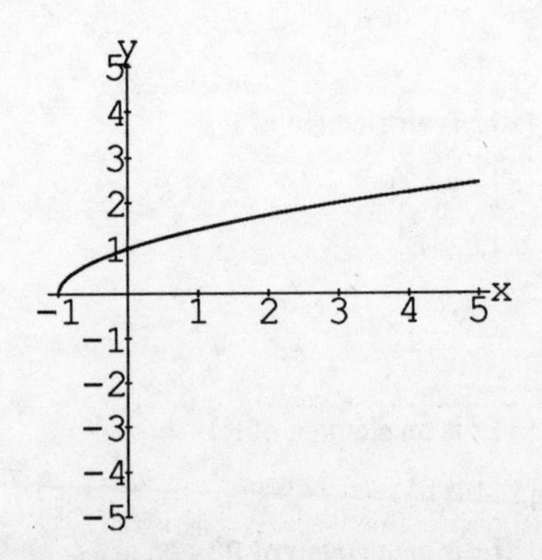

square root function

Index

E

F

G

H

I

J

L

M

N

O

P

Q

R

S